International Congress of Gynecology and Obstertrics

Shortening the Round Ligaments

Indications, Technics, and Results

International Congress of Gynecology and Obstertrics

Shortening the Round Ligaments
Indications, Technics, and Results

ISBN/EAN: 9783742860835

Manufactured in Europe, USA, Canada, Australia, Japa

Cover: Foto ©berggeist007 / pixelio.de

Manufactured and distributed by brebook publishing software
(www.brebook.com)

International Congress of Gynecology and Obstertrics

Shortening the Round Ligaments

[Reprinted from the American Gynæcological and Obstetrical Journal for December, 1896.]

SHORTENING THE ROUND LIGAMENTS; INDICATIONS, TECHNICS, AND RESULTS.*

By George M. Edebohls, A. M., M. D., New York,

Professor of Diseases of Women, New York Post-Graduate School; Gynæcologist, St. Francis Hospital; Consulting Gynæcologist, St. John's Hospital.

HISTORICAL.

Alquié, of Montpellier, France, appears to have been the genius who first conceived the idea and proposed a plan of shortening the round ligaments to correct downward and backward displacements of the uterus. In his historical memoir to the Académie de Médecine (1) he calls the operation *utero-inguinoraphie.* He had performed it only upon animals and the dead subject, never having operated upon the living woman. The proposition of Alquié was referred to a commission composed of Baudelocque, Bérard, and Villeneuve, who, after wrestling with the problem for four years, finally reported to the Académie (2) in substance as follows: They condemned the proposal of Alquié *in toto,* both as regards the possibility of permanently correcting uterine displacements by shortening the round ligaments and as regards the practicability of the operation itself. They asked the Académie, however, to pass a vote of approbation of Alquié for his prudence in never having attempted

---

* Read before the Second International Congress of Gynæcology and Obstetrics, Geneva, Switzerland, September, 1896.

his operation upon a living woman. The report was received by the Académie and acted upon in the sense of the reporters.

The first note of mild approbation comes from Aran (3), who, in advance of the time in which he lived, grasped the full significance of Alquié's proposition that shortening the round ligaments will bring into position a uterus displaced downward or backward, or both, but was deterred by the difficulties of the operation, which he considered one next to impossible of successful execution.

Deneffe (22), with the courage of the recent graduate, was the first to attempt the operation upon the living woman at Ghent, Belgium, in June, 1864. During his student days he had frequently practiced the operation successfully upon the cadaver, and, immediately upon obtaining his degree, requested of Burggraeve and Soupart an opportunity to shorten the round ligaments upon the living woman, avowing his confident belief that he would be able to carry the operation to a successful conclusion. His request was granted, and, in the presence of the masters, the operation was undertaken upon a patient suffering from prolapsus. Deneffe, however, failed to find either round ligament, although, following a suggestion of the great Burggraeve, he opened up the entire inguinal canal on both sides. The patient recovered from the attempt, but Burggraeve was promptly called to account by the Commission des Hospices for permitting experiments upon patients committed to his care.

Thereafter the idea seems to have slumbered in the medical mind, although Freund (4), and probably others, occasionally attempted the operation upon the cadaver, until William Alexander (5), of Liverpool, performed the first successful shortening of the round ligaments upon the living woman on December 14, 1881. Very soon after, in February, 1882, James A. Adams (6), who had for two years previously both practiced and taught the operation upon the cadaver, and recommended to his students its performance in suitable cases upon the living woman, performed his first operation—unsuccessful, however, on account of posterior adhesions of the uterus—upon a patient suffering from prolapse.

Such are what are believed to be the correct data in relation to the early history of the operation under discussion. From these beginnings the operation, although viewed askance in its infancy, and in spite of the fact that its performance to the majority of surgeons is not a congenial undertaking, has steadily, though slowly, gained favor and adherents as the ideal method of surgical procedure in those cases to which experience has shown it to be adapted. In America, probably, earlier and to a wider extent than in any other country, has it found due recognition; there also does it count its most numerous and enthusiastic supporters.

NOMENCLATURE.

Alquié himself entitled the operation *utero-inguinoraphie.* Only one writer (29) seems to have followed him in the use of this designation, which seems inappropriate in that neither the uterus nor the groin is sutured, but the essential of the operation is the resection and suture of the round ligaments. Cittadini (137) calls it *hysteropexie ligamentaire. Abbreviatio ligamentorum rotundorum uteri,* used by Riasentseff (50, 70, 71), itself needs abbreviation. It has been customary since 1882, when the first publications of Alexander and Adams called attention also to the original proposition of Alquié, to designate the operation variously, according to the writer's conception of the relative share of honor due each of these three men. Patriotism possibly, in some instances, also had more or less to do with the matter. Thus it has been called the Alquié, the Alexander, and occasionally the Adams operation; and two, and even all three, of these names have been combined in designating the procedure. Alquié was undoubtedly the genial originator of the idea. Alexander performed the first successful operation, and, in addition, deserves our thanks for the persistence with which he kept the operation before the profession by his frequent publications. Adams was preceded by Alquié in the conception of the operation, by Deneffe in its unsuccessful attempt upon the living woman, and by Alexander in date of operation, in successful execution, and in publication. The names of each of the three are indelibly and impressively associated with the operation, and can never be obliterated from its history. While second to no one in his admiration of the genius which originated the idea of·this ideal operation, as well as of the courage which first attempted and the success which first attended its performance, the writer believes that the time has arrived when names should be dropped in the designation of the operation, and when, until some briefer and just as expressive term be proposed, the operation should be known as shortening the round ligaments.

INDICATIONS.

From the outset the writer would have it clearly understood that the present paper deals exclusively with extraperitoneal or inguinal shortening of the round ligaments, as distinguished from intraperitoneal shortening of the round ligaments after either cœliotomy or vaginal section.

The superiority of shortening the round ligaments over its competitors, the various forms of ventral and vaginal fixation of the uterus, intra-abdominal shortening of the round ligaments, cystopexy of the uterus, and operative procedures on the utero-sacral ligaments, lies in the facts (*a*) that, while being equally successful and efficient, it is the most physiological of all these operations, both as regards its plan and the character of the results obtained, and (*b*) that it interferes to a less degree—indeed, not at all—with the functions proper of the uterus, childbearing and childbirth.

Comparison of Anatomical Result with that obtained by Other Operations.—Ad (*a*). The writer is quite free to admit that an anatomical cure of retroversion can be obtained by each of the above-named operations in from ninety to one hundred per cent. of cases by proper technics and a capable operator. Mackenrodt figures ten per cent. of failures for vaginal fixation. Kellogg, who has probably had the largest individual experience in shortening the round ligaments, writes me: " I have done the operation more than five hundred times, and have had failures in less than five per cent. of the cases." In ventral fixation there is no reason why the average operator should not obtain from ninety-five per cent. upward of anatomical cures.

The quality of the cure, however, is an entirely different matter, especially to the patient. After a successful shortening of the round ligaments the physiological mobility of the uterus remains unimpaired, *no peritoneal adhesions having been established.* In future pregnancies the shortened round ligaments undergo evolution and involution with the uterus. Essential to *all* of the rival operations is the establishment of a condition *always* pathological, *never* physiological, the deliberate creation, repugnant to every surgical instinct, of more or less extensive, more or less firm, peritoneal adhesions. On this score alone, I contend, shortening of the round ligaments, whenever applicable to the particular case, should receive the preference of the true surgeon.

Comparison of the Course of Subsequent Pregnancies.—Ad (*b*). Their greater or less interference with the functions proper of the uterus—childbearing and childbirth—constitutes the higher standard by which these various operations must now be judged. The writer (186) has recently gone somewhat extensively into this subject, and will not reiterate details here. The record of disasters of pregnancy and parturition following the yet young operation of vaginal fixation of the uterus is already so appalling as to justify the dictum that it has no place, or rather is absolutely contraindicated, in the case of any woman with the possibility of a future pregnancy before her.

Milaender (*Zeitschr. f. Geb. u. Gyn.*, vol. xxxiii, No. 3) and Noble (193) have collected a large number of cases of pregnancy following ventral fixation of the uterus. While the record of ventral fixation in this respect is not quite so bad as that of vaginal fixation, it is still bad enough to impose on us the duty of limiting to as great an extent as possible the field of ventral fixation of the uterus.

Contrast with this the course of pregnancy and parturition after shortening the round ligaments. A slight drawing pain, beginning with the eighth month of pregnancy, and attributed, whether correctly or not, to traction upon the shortened ligaments, has been reported in a few cases. Beyond this, disturbances of pregnancy or parturition, *due in any way to the operation*, have not been observed. This is all the more remarkable when we consider the length of time during which the operation has been before the profession and the numerous subsequent pregnancies which have been reported by many writers. Foreman (61), as long ago as 1887, reported seventeen pregnancies in sixty-seven cases in which he had shortened the round ligaments, all of which terminated by easy deliveries at term. Gardner (41), Alexander (54), Batchelor (151), Johnson (188), and others too numerous to mention, testify to the uniformly smooth and normal course of subsequent pregnancies and labors, recording only very exceptional cases of abortion. The writer's cases of pregnancy following shortening of the round ligaments will be detailed under the head of results.

Broad General Indication and Corollaries.—I have dwelt thus long upon the distinguishing features of the operation of shortening the round ligaments as compared with its rivals—viz., (*a*) preservation of the physiological mobility of the uterus; (*b*) non-dependence for success upon the establishment of intraperitoneal adhesions;

(c) the equal permanence, as well as the better character, of its anatomical results; (d) non-interference with the course of future pregnancy and parturition—since a consideration of these features of superiority leads unavoidably to the establishment of one grand indication to which all other indications are subsidiary or corollary. This dominant law might be formulated as follows: *Shortening of the round ligaments is indicated whenever and wherever it will meet the indications as well as or better than one of its rival procedures, i. e.:*

1. In all uncomplicated cases of retroversion, retroflexion, and excessive mobility of the uterus requiring operative treatment.

2. In cases of aggravated anteflexion of the uterus, when the fundus is below the level of the internal inguinal ring—*i. e.*, in all uncomplicated anteflexions worthy of the name.

3. In cases of retroverted anteflexed uteri without adhesions.

4. In simple prolapse of the ovaries without adhesions, when that condition calls for treatment.

5. In cases of adherent retrodisplaced uteri, with or without adhesions of tubes and ovaries, these organs being otherwise in condition to call for an attempt at their conservation. The adhesions in these cases are first to be separated, at the preference of the operator, by means of anterior or posterior colpotomy, median cœliotomy, or by an incision through the peritonæum at the internal inguinal ring, and enlargement of the latter to permit the introduction of one or two fingers. The writer has practiced each of these procedures, the last of which he believes to be original with him.

In two instances I have shortened the round ligaments in order to enable the patients to dispense with the use of pessaries, which they were compelled to wear to maintain the uterus in anteversion. Both ladies considered the wearing of a pessary and all that it implies a nuisance, deliverance from which was not purchased too dearly at the cost of an operation.

Shortening the Round Ligaments not the best Prolapsus Operation. ·-The above appear to the writer to be all absolutely indisputable indications for shortening the round ligaments. In addition, the operation has been practiced by its originators, and by nearly all of their followers, for the cure of prolapsus uteri. When practiced in these cases it should be clearly understood that shortening the round ligaments is but an adjuvant operation to the necessary plastic work upon uterus, vaginal walls, and perinæum called for by the conditions presenting in each case, and that, in order to secure the best results, *all* operations called for should be performed at the same sitting. Although among those who practiced this method with success, the writer early became convinced that the function of the round ligaments is not to sustain the uterus from dropping out of the pelvis, and that whatever success had been reached in the treatment of prolapsus uteri by the combination of shortening the round ligaments with plastic work, was due in greater part to the plastic work, and in a minor degree to the anteversion secured by shortening the round ligaments. Since December, 1890, in place of shortening the round ligaments, the writer has employed ventral fixation of the uterus, performed at the same sitting with the plastic work called for by each case, in the operative treatment of complete prolapsus uteri et vaginæ. This combination of ventral fixation and plastic work was taken up some three years later by Kuestner. German writers, either overlooking or ignoring the author's prior work and publication (Edebohls, The Operative Treatment of Complete Prolapsus Uteri et Vaginæ, *Am. Jour. Obst.*, vol. xxviii, No. 1, 1893, and Combined Gynæcological Operations, *Am. Jour. Med. Sciences*, September, 1892), call it Kuestner's method.

OBJECTIONS.

Of the objections to the operation of shortening the round ligaments the greater part, especially that relating to the alleged more or less frequent absence of the ligaments, are purely theoretical (Gehrung (12), Smith (52), Bird (152), and others), the more tangible ones concerning themselves chiefly with the technics. These objections have been so frequently, so fully, and so ably answered by others—notably by Kellogg (161)—that I will refrain from entering upon them here. The only serious objections that can be justly urged are the occasional occurrence of hernia and of pains in the region of the scar. How the risk of these occurrences may be minimized, or possibly entirely obviated, the writer will endeavor to show under the head of Technics.

Abnormal Course and Insertion of Round Ligaments.—Regarding the alleged more or less frequent absence of the round ligament, the writer would merely add his testimony to that of other competent observers that the round ligament is a structure constantly present. He failed to find it in one instance only (Case XVI), but the fault lay with his insufficient development as an operator upon the roun' ligaments. He hopes to have atoned in some measure for this shortcoming by subsequently finding four round ligaments (the right in Case LVI and Case XCVIII, and both ligaments in Case XCII) that were not in their normal place in the canal. So far as my reading and knowledge go these are the first and only instances in which the round ligaments have been demonstrated to be present in full development while at the same time absent from their normal place in the inguinal canal. The round ligament in each of these instances ran its usual course from the cornu of the uterus to the internal abdominal ring, but, immediately on leaving the latter, instead of running downward and inward in the inguinal canal, it turned abruptly to run upward and outward, behind the transversalis muscle, to be inserted into the outer half of Poupart's ligament and the adjacent portions of the transversalis fascia. Three of the ligaments were found only after tracing them, through a median incision, from the cornu of the uterus to the internal ring, and thence onward in their erratic course to their abnormal insertion, *none* of the fibers, be it distinctly understood, running downward and inward along the canal. The fourth abnormal ligament was found and traced from the internal ring upward and outward by enlarging the external incision. Each of the four abnormal ligaments was cut away from its insertion above, shortened, brought down into its normal place in the canal, and sutured in the usual way (175). Since the above was written, my house surgeon at St. Francis Hospital, Dr. T. A. Lehmann, in shortening the round ligaments of one of my hospital patients, found the ligament of the right side taking the same abnormal course and having the same abnormal insertion as in the four cases just detailed. He successfully followed the same tactics in the management of this case that were employed in three of my operations.

The explanation of this anomaly is not so very clear, but the writer would hazard the surmise that its ætiology is identical with that of non-descent of the testis and spermatic cord in the male, the round ligament being the analogue of the spermatic cord.

The Operation not Congenial to Every Surgeon.—A contra-indication to the operation of shortening the round ligaments that may be allowed by some is inability of the operator to perform the operation. Others, again, will not admit this, and contend that in that case the surgeon should either learn how to do the operation or send his patient to some one who can perform it successfully. The operation is certainly one that is not congenial to every surgeon; the majority of operators will always find it easier to find the fundus uteri through an abdominal or vaginal incision and stitch it forward than to find, isolate, and shorten the round ligaments in the inguinal canal. It is this difficulty, relatively speaking, of the operation of shortening the round ligaments which has stood in the way of its earlier and more general adoption. Yet, though Adams, one of the originators of the operation, said of it, " The operation is one that all and sundry can not perform," the writer is convinced that, with a knowledge of anatomy and a little practice upon the cadaver, any surgeon capable of doing a Bassini operation for the radical cure of inguinal hernia should be able to shorten the round ligaments successfully.

TECHNICS.

General Considerations.—The essentials of shortening the round ligaments successfully are: 1. To find the round ligaments. 2. To isolate, draw out, and sufficiently shorten them. 3. To properly anchor the external ends of the shortened ligaments.

To begin with, operators have preferences as to which side of the patient to stand on while operating; a few even change sides during the course of the operation. An occasional one stands between the thighs of the patient. The writer prefers to stand on the right side of his patient and to begin the operation upon the left ligament. Next, as to the external incision, the vast majority of operators prefer to incise parallel to Poupart's ligament, the lower end of the incision generally corresponding to the external inguinal ring. The length of incision varies from the neat two- to four-centimetre cuts of Alquié (2), Imlach (27), Kellogg (86), Newman (91), Cleveland (174), to the ghastly gash of Kocher (146), who lays open skin and fat from pubis to anterior superior spine of ilium. Foreman (61) makes a ⌐-shaped incision for æsthetical reasons; Duret (143), a curved incision, with the convexity downward, from one external ring to the other. Alquié, Kellogg, and Newman incise over the middle of Poupart's ligament.

Finding the Ligament.—In searching for the ligament three principal methods have been adopted. The first is to seek and draw out the round ligament at the external ring. Alexander (5) operated in this way upon his first case, and the great majority of all operators since have followed him in this method. The second method, that originally proposed and practiced upon the cadaver by Alquié himself, is to incise or puncture the anterior wall of the inguinal canal higher up, near the internal ring, and through this small opening to hook up the ligament in the canal and draw it out. Kellogg (86) was the first to adopt this as a routine procedure. Newman (91), following a suggestion of Frank, punctured the anterior wall of the canal a little higher up than Kellogg, and claimed priority. His method, as far as can be gathered from his very deficient and imperfect first description (91), differed in no essential or principle from that of Kellogg, except that Kellogg punctured the anterior wall of the canal near its middle, Newman a trifle higher up—a distinction without much of a difference. Newman's (91) first case was operated upon some three months after Kellogg read a paper (106, page 17 of reprint) reporting a number of operations performed by his new method.

The writer (123) was the first to propose incision of the entire length of the anterior wall of the canal *as a routine procedure* in the operation of shortening the round ligaments. Others before him had, for various reasons, *in exceptional cases*, incised the anterior wall of the canal to a greater or less extent. Deneffe (22), in the very first operation attempted upon the living woman, after failing to produce the round ligaments through the incision of Alquié, at the suggestion of Burggraeve opened the entire inguinal canal in the vain attempt to find them. Reid (15), after breaking a round ligament, incised the canal for half an inch to aid him in again finding it. Alexander (17) himself, in one of his later papers, refers to the occasional necessity of partially opening the canal. Roux (95) derived help from incising the arciform fibers to the extent of one centimetre. Blake (134) opened the canal on the left side, after experiencing great difficulty in finding the right round ligament at the external ring of the same patient. A number of operators— Chalot (136), Kocher (146), Werth (171), Kuestner (178), Fabricius (176)—have, in the course of years, followed the writer in advocating incision of the entire length of the anterior wall of the canal.

The name of Newman has, under misapprehension, been so often associated with the writer's (123) modification of the operation of shortening the round ligaments, originally proposed in 1890, on account of the hasty, ill-considered, and baseless claim of Newman (132, 130), that this would appear to be the proper place to empha-

size the fact that Newman's technics come into competition with
the prior method of Alquié-Kellogg (2, 85, 88, 106) rather than
with those of the writer. In other words, there is nothing essentially
original in the operation described by Newman (91).

Isolating and drawing out the Round Ligament.—To find, isolate,
and draw out the round ligament is not always an easy matter.
The difficulties are materially increased when the ligament is sought
and drawn out at the external ring instead of higher up in the inguinal canal. This is due to the anatomical structure of the round
ligament, which runs as a single cord from the cornu of the uterus
to and through the internal ring and the upper portion of the inguinal canal. It divides up and its anatomy becomes more complicated, however, just within and at the external ring. Adams
(6, 183), himself an expert anatomist, quotes Rainey (*Trans. Royal
Soc.*, 1850) as giving the most complete and altogether reliable description of the round ligament: " The so-called round ligaments
of the uterus, regarded as a muscle, may be said to arise by three
fasciculi of tendinous fibers—the inner from the tendon of the internal oblique and the transversalis, near to the symphysis pubis;
the middle, from the superior column of the external ring near to
its upper part; and the external fasciculus from the inferior column
of the ring just above Gimbernat's ligament. From these attachments the fibers pass backward and outward, soon becoming fleshy.
They then unite into a rounded cord, which crosses in front of the
spermatic artery. . . . It then gets between the two layers of the
peritonæum forming the broad ligament, along which it passes
backward, downward, and inward to the anterior and superior part
of the uterus, into which its fibers, after spreading out a little, may
be said to be inserted." From a most minute and brilliant practical
demonstration of the anatomy of the round ligaments by Professor
James E. Kelly, of New York, which I had the privilege and pleas-
ure of witnessing, I am convinced that the above description is
practically correct. When operating at the external ring the difficulties of collecting these scattered fibers of the round ligaments at
and near the ring and tracing them upward to where they unite to
form a single cord, are additional to those necessarily encountered
in separating the round ligament from the ilio-inguinal nerve and
fat, as well as from the muscular and tendinous fibers of the internal
oblique, which—analogues of the cremaster muscle in the male—
invest the round ligament in the canal. When, in addition to this,

it is considered that the round ligament becomes stronger, and is less likely to tear in manipulation, higher up in the canal than at or near the external ring, the advantages, apart from other considerations, of picking it up within the canal rather than at the external ring become apparent.

In drawing out the round ligament the accompanying ilioinguinal nerve should be carefully separated from the ligament and guarded against division. This will prevent the subsequent pains in and about the cicatrices which are frequently observed when this nerve is divided. The separation of nerve and ligament is difficult only in the canal, where they run side by side. At the upper end of the canal their courses diverge, the ligament penetrating the internal ring, while the nerve continues upward and outward between the muscular layers of the abdominal wall. Werth (171) resects a portion of the nerve to prevent after-pains. With Kuestner (178), however, I have never met with subsequent pain when the nerve is stripped off and preserved.

Another point, to which the writer believes he was the first to call attention, is the advisability of stripping back the investing peritonæum of the broad ligament for a certain distance from off the round ligament in every case. This stripping back of the peritonæum, which is accomplished by the fingers aided by the sense of sight, has two things to commend it. Firstly, it avoids the danger, when drawing down the round ligament, of invagination of a peritoneal pouch or process into the canal, with the possible resultant invitation to descent of a hernia. Secondly, we are assured by this measure that the round ligament is really shortened in its intra-abdominal course between the cornu of the uterus and the internal ring, and not merely stretched in the canal, which, of course, would have no effect upon the position of the uterus. Neglect to shorten the round ligaments *sufficiently*, by stripping back the investing peritonæum, explains many a case of failure or semifailure of what otherwise would have been a successful operation, as well as the necessity for the use of pessaries after operations by some operators. The corrugated cuff of stripped-back peritonæum, in addition, forms a very effective plug at the internal ring, an additional safeguard against hernia. Newman (132) and Adams (183), in his last paper, also advocate this procedure. In practicing it the writer draws out the round ligament and strips back the peritonæum until the finger passed down to the internal ring recognizes the impact of the cornu uteri of the same side as it is drawn forward by the round ligament.

The amount of shortening of the round ligaments necessary to be accomplished averages about ten centimetres, and, to make an assuredly successful operation, should not be less than seven centimetres. Alexander (7) draws out the round ligament farther when operating for prolapsus than when dealing with retroversion.

Anchoring the Shortened Round Ligaments.—In no other part of the operation of shortening the round ligaments, or perhaps of any other operation, has the same amount of ingenuity and fertility of resource been displayed as in the attachment and suture of the shortened ligaments. Of those who operate upon and draw out the ligament at the external ring, nearly all stitch the shortened ligament to the pillars of the ring. Alexander (7), in addition, stitches the end of the ligament to the margins of the skin wound, Lee (87) to the periosteum, and Roux (95) to the spine of the pubis. Cleveland (174) brings the ligament out through a puncture of the skin at a little distance from the operation wound, and secures it there with a suture. Carpenter (120) cuts away the excess of ligament an inch or so outside of the ring, splits this inch of ligament longitudinally, and sews each half to the external fascia. In disposing of the excess of ligament the above-named operators, in the majority of instances, simply cut it away. Foreman (61), as a rule, and Adams (183), sometimes, stow away the slack in the wound. Abbe (*New York Med. Jour.*, March 17, 1888) draws one ligament subcutaneously through the fat overlying the pubis across to the opposite side, and there ties it in a living knot with its fellow. Doleris (102), Batchelor (151), and Martin (189) follow Abbe's plan with insignificant modifications. Duret (143) also accepts the same principle, fastening the two ligaments to each other, however, at the bottom of a crescentic wound extending across from one external ring to the other.

Kellogg (67, 106), who, as already stated, draws out the ligament through a small opening in the anterior wall at the upper part of the canal, has varied his method of securing the ligament from time to time. Originally he secured the ends of the ligaments with silver wires, which he tied together over a hard-rubber plate placed on the skin between the wounds. Subsequently he tucked the excess of ligament down into the lower part of the canal. Still later, he drew out the slack of ligament through a separate small puncture lower down in the aponeurosis of the external oblique, leaving the ligament attached at the external ring, and folding the excess in the fat of the external wound.

Of those who prefer to open the inguinal canal along its entire length, Chalot (136), Kuestner (178), and the writer (123) in his first operations, fastened the shortened round ligaments within the canal just behind the aponeurosis of the external oblique, the cut margins of which were brought together by a series of sutures which also pierced and secured the ligament. Kocher (146) turns the drawn-out ligament upward at the upper and outer angle of the fascial wound, and sews it on to the outer surface of the external oblique aponeurosis in the direction of the anterior superior spine of the ilium. Fabricius (176) and the writer (186, 187), in his later cases, place and secure the shortened ligament in its natural habitat behind the lower border of the internal oblique.

The Accident of tearing the Round Ligament.—When a round ligament tears in drawing it out, what shall we do? The answer to this will depend upon the site of the tear. If the ligament has parted within the canal, the retracted end should be sought for and recovered, if possible, at the internal ring. If this should prove impossible, the peritonæum should be incised at the internal ring and the ligament sought for and brought out of the abdomen with the aid of a finger or two, assisted, if necessary, by slender forceps. If the torn end of the ligament can not be readily found and grasped by the forceps, the finger should be hooked behind the broad ligament of the corresponding side and passed along the posterior surface of the latter until the cornu uteri is reached. The uterus, and with it the torn ligament, is then lifted forward and the round ligament traced from the cornu uteri outward to its torn end, which is then grasped by forceps, brought out, shortened to the requisite degree, and sutured in the usual manner.

The same course may be successful if the round ligament should happen to tear just within the internal ring. In the only instance in which this happened to me I made a small incision in the median line of the abdomen, sought the ligament at its uterine end, traced it outward to the tear, pushed the torn end out of the abdomen through the internal ring, and successfully completed the operation.

When the round ligament tears out of the uterus, or so near the latter as to leave a stump too short to be drawn out and properly fastened, three courses are open to us. The first is to shorten the opposite ligament, and trust to one shortened ligament to keep the uterus forward. This course has been followed by success in a case reported by Ledyard (13), and the writer has learned of several other successes from personal communication with the operators. Gottschalk (*Ges. f. Geb. u. Gyn.*, Berlin, November 8, 1889) has reported bad effects resulting in a case in which only one round ligament was shortened. In this connection it is interesting to note that Blake (134) advocates operating on one ligament only at a time as a routine procedure, and that Brown (77) quotes Doleris as advising the same procedure in certain cases in order to allow the bladder room for expansion on the opposite side prior to shortening the second ligament.

The second alternative, originally suggested by the writer, but thus far practiced neither by himself nor by any one else to his knowledge, is to open the abdomen in the median line, unite the torn ends of the round ligament, or attach the pulled-out end to the uterus, by suture, and complete the operation in the usual way. Our third resource is to substitute immediately the next best retroversion operation. The writer followed this course in five instances in which the round ligament parted at or near the uterus, performing ventral fixation at the same sitting, with resultant cure of the retroversion in each case.

Strength of Round Ligament.—In estimating the amount of traction the round ligaments will bear without tearing, experience will probably prove to be our main help. In a large number of experiments made by the writer, both on the dead subject and with pieces of round ligament obtained at operation, the force necessary to tear the round ligament varied between the limits of three and fifteen kilogrammes.

The Wearing of Pessaries after Operation.—There exists among operators some divergence of view, both as to the best method of anteverting the uterus before or during operation and as to the necessity of artificial support of the uterus after operation. Some recommend and some condemn the use of the sound to replace the uterus before shortening the round ligaments. Some consider the wearing of a pessary in cases of retroversion or of an intrauterine stem in cases of retroflexion for a certain variable period after operation as well-nigh indispensable to success. My own practice in these regards has been as follows: After the patient is under ether I satisfy myself that the retroverted uterus can be brought into pronounced anteversion by bimanual manipulation; that tubes and ovaries are normal in size and not adherent. I then curette the uterus, allow it to assume any position that it will, and proceed to shorten the round ligaments, lifting the fundus forward during the operation by traction on the ligaments. I have never had my patients wear any kind of support after operation. The positive shortening of the intra-abdominal portion of the round ligaments accomplished by the technics which I employ renders a pessary or support after operation entirely superfluous.

The Question of Hernia following Operation.—From the very beginning the possibility of hernia following the operation of shortening the round ligaments has been ever present in the minds of operators and measures of precaution against this accident have entered into the technics of most surgeons. Previous to the advent or just appreciation of the value of the Bassini operation for the radical cure of inguinal hernia these precautionary measures were of a nature as unsatisfying as were the then known operative procedures against hernia themselves. Not that hernia in any considerable degree of frequency followed the operation of shortening the round ligaments; yet the slight risk of its occurrence formed the chief and perhaps only well-founded objection to the operation. Among others, Foreman (61) early included the deep abdominal muscles in his sutures, burying the latter. Fry (124) mistakenly advocated open treatment of the wound to prevent hernia. Kocher (146), after drawing out the ligament and securing it as already described, closed the canal by a row of sutures embracing the entire thickness of the muscles of the abdominal wall, and tied upon the fascia of the external oblique. Yet these measures were all more or less crude, and not up to the standard of modern surgery, until Fabricius (176) and the writer (186, 187), each in his own way, *correctly* applied the essential principle of the Bassini operation—the sewing of the lower edge of the internal oblique and transversalis muscles to Poupart's ligament, and the obliteration of the inguinal canal—to the technics of shortening the round ligaments. To Fabricius (176), between us, belongs priority both of execution and of publication. He unites the internal oblique and transversalis to Poupart's ligament by a row of interrupted buried-silk sutures, each of which also pierces the round ligament. Over this he closes the anterior wall of the canal by a second row of interrupted silk sutures.

The technics employed by the writer will be described in detail in its proper place. The question of hernia will be again referred to in the analysis of the author's statistics.

Suture Material and Drainage.—In the matter of suture material and drainage the same latitude of opinion and practice obtain as in other departments of operative surgery. The writer has tried every form of suture material, but now uses exclusively the buried running suture of forty-day catgut for the deep parts and a subcutaneous catgut suture to unite the skin. In the first half of my cases I drained the deep parts of the wound with a few strands of silkworm gut; the last sixty or so have, with a rare exception now and then, not been drained.

Anæsthesia.—Shortening the round ligaments can be performed, if necessary, without general anæsthesia. Kellogg (106) operates quite frequently under local cocaine anæsthesia. Ether has been the anæsthetic in nearly all of my cases; I have, however, operated under cocaine, with Schleich's infiltration anæsthesia, and without any anæsthetic whatsoever.

Comparison of Different Methods of shortening the Round Ligaments.—Before entering upon the description of my present method of shortening the round ligaments, it is but proper to state that I have had personal experience with each of the three principal methods under discussion—that of operating at the external ring (Alexander), that of hooking up the ligament through a puncture of the anterior wall of the canal near the internal ring (Alquié-Kellogg), and that of laying open the canal along its entire length (Edebohls).

After operating upon seventy-four cases after his (123) original method, the writer, in September, 1894, during a visit to Battle Creek, Mich., saw Dr. Kellogg shorten the round ligaments of two patients. I was so fascinated by the smallness of the incision required, by the neatness of the entire operative technics, by the dexterity and celerity (nine minutes exactly) with which Dr. Kellogg performed his operation, that I at once discarded my own method and adopted Kellogg's. I succeeded with it quite satisfactorily, in a few instances, however, having to open the canal along its entire length to find, draw out, and secure the ligament. In the twenty-six cases in which I practiced Kellogg's method three ligaments (Cases XCII and XCVIII) were unfortunately not present in the canal. In some of the other cases other tissues—in one instance even the femoral artery—were hooked up and drawn out before the ligament was found. I should have persisted in a further trial of the method, however, and perhaps have adopted it permanently, had I not missed too greatly the ease and certainty with which the ligament could be found and isolated after opening the canal, as well as the welcome test of the proper amount of shortening afforded by passing the finger to the bottom of the canal and feeling the cornu uteri at the internal ring when the ligament was drawn taut. The only thing I did not quite like in my old operation was the method of securing the ligament just behind the aponeurosis of the external oblique. With the modification, however, of closing the wound by a typical Bassini and of suturing the ligament in its natural habitat, this objection, together with that of the risk of hernia, seemed to be overcome. Since January, 1896, when he first adopted it, the writer has practiced his present method in fourteen cases.

To compare once more and briefly the three methods of shortening the round ligaments, let us see how each of them meets the three requirements of a successful operation: (1) To find the ligaments, (2) to isolate and draw them out sufficiently, (3) to properly anchor the shortened ligaments. It is self-evident that the round ligaments can be more readily found and more easily, directly, and gently separated from their surroundings after the entire canal is opened than by fishing for and drawing them out either at the external ring or through a small opening in the anterior wall of the canal near the internal ring. Moreover, in cases of abnormal course and insertion of the round ligament, five instances of which have been detailed above, the real condition present could be known with certainty only after opening the entire canal.

In regard to the third postulate—the proper anchorage of the shortened round ligaments—none of the many methods already referred to compares in directness and naturalness with that of anchoring the round ligaments in their normal situation in the canal. This can be done properly in but one way—*i. e.*, with the canal widely open.

AUTHOR'S PRESENT METHOD OF SHORTENING THE ROUND
LIGAMENTS.

On the day preceding operation the bowels are emptied by a laxative, the pubis and external genitalia are shaved, and a cleansing bath is given. Eight to twelve hours preceding operation an ointment, composed of creolin (ten parts) and mollin (ninety parts), is rubbed into the skin of the lower abdomen, pubes, and adjacent surfaces of the thighs. The parts are then covered with sterile gauze and left undisturbed until the patient is upon the table and anæsthetized.

Just prior to shortening the round ligaments the uterus is always curetted, and whatever plastic work upon cervix, vagina, and perinæum the conditions presenting in each case call for is performed. If adhesions of the uterus and annexa exist, and the operator prefers to sever these adhesions by anterior or posterior colpotomy rather than by an incision from above, this is the proper time to do so. At all events, the operator must satisfy himself that the uterus can be well anteverted *by bimanual manipulation* before proceeding with the operation of shortening the round ligaments. The uterus is then allowed to assume any position it may please, generally dropping backward, to be brought into position at a later stage of the operation by traction on the round ligaments. A little iodoform gauze is loosely placed in the vagina, not to sustain the uterus, but as an antiseptic precaution in view of the preceding curettage. The field of operation is lathered and scrubbed with more ten-per-cent. creolin-mollin, rinsed clean with sublimate solution (1 to 3,000), and the patient is ready for operation. In shortening the round ligaments I prefer to have the pelvis slightly elevated and to stand at the right side of my patient, beginning the operation upon the left ligament.

An incision five to six centimetres long, and nearly parallel to Poupart's ligament, is carried from the site of the internal inguinal ring downward and inward, terminating just within the spine of the pubis. Careful location of the pubic spine, from the time of beginning the operation until the anterior wall of the inguinal canal is opened, is absolutely essential to success. The subcutaneous fat is divided until the glistening aponeurosis of the external oblique muscle is exposed. The superficial epigastric artery is frequently divided, and if so should be ligated in this stage of the operation. The external inguinal ring is now either exposed to view or located by the touch. A grooved director is inserted through the external ring and passed along the inguinal canal, directly behind the aponeurosis of the external oblique, until its point is over the site of the internal ring. Cutting upon the director exactly in the direction of the fibers of the external oblique aponeurosis, one sweep of the knife lays open the anterior wall of the inguinal canal along its whole length (Fig. 1).

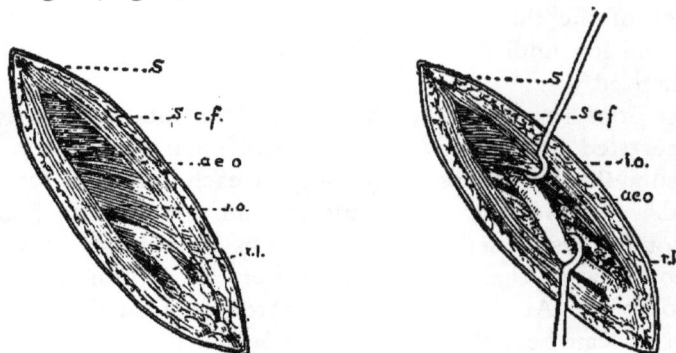

FIG. 1.   FIG. 2.

FIG. 1.—Incision, 5 centimetres long, through aponeurosis of external oblique, laying open inguinal canal from external to internal ring and exposing internal oblique muscle and round ligament. The ligament is more or less concealed according to greater or less development of internal oblique. *S.*, skin; *s. c. f.*, subcutaneous fat; *a. e. o.*, aponeurosis of external oblique; *i. o.*, internal oblique; *r. l.*, round ligament.

FIG. 2.—Isolating round ligament from its attachments in inguinal canal. *S.*, skin; *s. c. f.*, subcutaneous fat; *i. o.*, internal oblique; *a. e. o.*, aponeurosis of external oblique; *r. l.*, round ligament.

It is very desirable that all hæmorrhage should be controlled before opening the inguinal canal, otherwise the flow of blood into the latter may render differentiation of the round ligament from the other contents of the canal exceedingly difficult. An assistant exposes the contents of the canal by drawing apart the lips of the incision through the external oblique aponeurosis, with the aid of tenacula, blunt hooks, or clamp forceps. The lower fibers of the internal oblique muscle are seen crossing the upper half of the canal, filling it more or less, according to the greater or less muscular development of the individual.

In a fair proportion of cases the lower end of the round ligament is at once exposed to view, emerging from beneath the lower border of the internal oblique; more generally, the round ligament is well covered and entirely hidden from view by the internal oblique muscle and an investment of fatty, areolar, and fibrous tissue. Quite frequently some of the fibers of the round ligament are so closely interlaced with those of the internal oblique muscle that differentiation and separation of the ligament from bundles of muscular fiber becomes difficult. It is this part of the operation which generally trips the beginner; he fails to find the ligament, and can not, of course, proceed. Experience has taught me that the best method of procedure at this stage, if the ligaments are not at once exposed to view and recognized, is to search for them in the following manner (Fig. 2): Retract the internal oblique muscle upward and inward by a blunt hook passed beneath its lowermost fibers, and hand this hook to your assistant. Take two small blunt hooks, one in either hand, and sweep one of them, point downward and outward, along the posterior and outer walls of the canal from the depths of the wound skinward, hooking up the entire contents of the canal. By teasing these contents apart more or less, as required, by means of the two blunt hooks, the round ligament, surrounded by fat and muscular and tendinous fibers from the internal oblique, and accompanied by the ilio-inguinal nerve, will soon be recognized, and can be followed along the canal to the internal ring. There the round ligament is always strong, however weak, thin, and frayed-out it may have been found lower down in the canal or at the external ring.

The ligament is next separated from its investments in the canal, leaving, however, the pubic end attached for the present. In this part of the operation great care should be exercised not to divide or tear the ilio-inguinal nerve which accompanies the ligament, and division of which is the cause of the various dysæsthesiæ in the vicinity of the scar sometimes complained of by patients after operation. In the canal itself the ilio-inguinal nerve and the round ligament are very intimately connected; at the upper end of the canal they diverge, the nerve to pass between the muscular layers and the ligament to enter the internal ring.

The ligament, freed from its surroundings in the canal, is next grasped by the thumb and forefinger of the right hand and cautiously drawn out at the internal ring (Fig. 3). The line of traction should be more or less perpendicular to the surface of the abdomen at that point, approximately in the direction of the intra-abdominal portion of the ligament. As the round ligament emerges at the internal ring it is seen to carry with it, in the form of an inverted cone, the investing peritonæum of the broad ligament, the point of reflection of the latter being marked by a distinct white line surrounding the round ligament. With the thumb and forefinger of the left hand the investing peritonæum is stripped or milked back into the abdomen as the round ligament emerges farther and farther from the internal ring. Occasionally the peritonæum tears in stripping it back; this is a matter of no consequence provided the asepsis is all it should be.

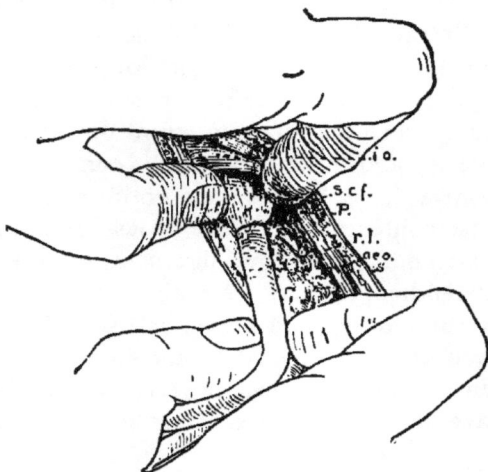

FIG. 3.—Drawing round ligament out of abdomen and stripping back investing peritonæum of broad ligament. *i. o.*, internal oblique ; *s. c. f.*, subcutaneous fat ; *P.,* peritonæum ; *r. l.*, round ligament ; *a. e. o.*, aponeurosis of external oblique ; *S.*, skin.

Should the ligament not run freely out of the abdomen, it will be wise, before employing the limit of safe traction force, to ascertain the cause by incising the peritonæum at the internal ring, bluntly dilating the latter, and passing a finger into the abdomen. If posterior adhesions prevent the uterus, tubes, and ovaries from coming freely forward, these may be separated by a finger or two hooked behind the broad ligament; or if the infundibulo-pelvic ligament, as obtained in one of the writer's cases, be shortened and thickened as the result of previous inflammation, this ligament may be stretched. The round ligaments will then be found to run freely, and the process of stripping back the peritonæum is continued until the index finger, passed down to the bottom of the wound, recognizes the impact of the cornu uteri at the internal ring when traction is made upon the round ligament. This constitutes the writer's index to the proper amount of shortening, which, expressed in figures, will average about ten centimetres.

The opposite round ligament is now sought, isolated, and drawn out in the same way. When free play of both ligaments has been secured, the anteposed fundus of the uterus may be drawn from side to side by alternating traction upon the ligaments, the movements of the fundus being recognized by the operator's fingers placed on the abdomen immediately above the pubis. In thin persons the transits of the uterus are frequently visible to the eye.

The proper course to pursue in case the round ligament should tear either in the canal or within the abdomen, as well as in cases of absence of the round ligament from the canal, has already been indicated.

Our next care is to properly reanchor the ligaments and close the wound. Thus far the ligaments have remained attached at their outer or pubic ends. These attachments are now cut for convenience in further manipulation, without, however, amputating any part of the ligament at present. After securing the desired position of the uterus by traction upon the round ligaments, and adjusting the latter nicely along the bottom of the canal, suture of the wound is in order.

The writer's suture material for the deep parts consists of catgut No. o, chromicized to resist absorption for about six weeks. Those who may be interested in the method of preparation and sterilization of forty-day catgut are referred to an article by the writer in the *American Gyn. and Obst. Jour.*, May, 1896, entitled What is the Best Method of making and of closing the Cœliotomy Incision? A half-metre length of this forty-day catgut is threaded upon a full-curved Hagedorn needle of medium size or under. An assistant, with two tenacula, holds wide open the lips of the incision through the aponeurosis of the external oblique, so as to clearly expose the deep parts of the canal, and especially the clean-cut projecting shelf of Poupart's ligament.

The parts are brought together after the principle of Bassini's operation for the radical cure of inguinal hernia, with the exception that, instead of the interrupted suture, the buried running suture of forty-day catgut, applied according to the following technics, is used: Beginning at the upper angle and inner side of the right wound, the first sweep of the needle pierces the aponeurosis of the external oblique, the underlying internal oblique and transversalis muscles, the margins of the internal ring, the round ligament as it emerges between them, and the projecting shelf of Poupart's ligament. The succeeding loops of the deep tier of sutures, three or four in number, pierce the internal oblique and transversalis muscles, the round ligament, and Poupart's ligament. The last loop, in addition, penetrates the outer pillar of the external ring, and emerges upon the outer surface of the external oblique aponeurosis

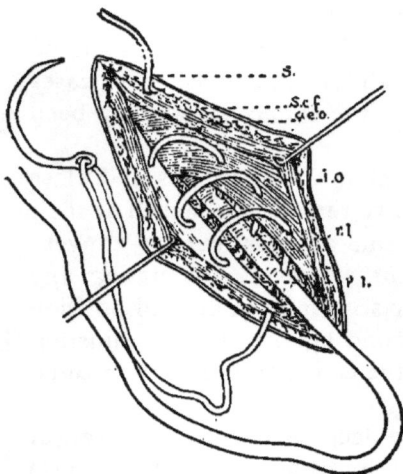

FIG. 4.   FIG. 5.

FIG. 4.—Deep tier of buried running suture of forty-day catgut, embracing internal oblique and transversalis muscles, round ligament and Poupart's ligament. Deep part of uppermost loop of suture (not showing in cut) passes at level of and embraces margins of internal ring. *S.*, skin; *s. c. f.*, subcutaneous fat; *a. e. o.*, aponeurosis of external oblique; *i. o.*, internal oblique; *r. l.*, round ligament; *P. l.*, Poupart's ligament.

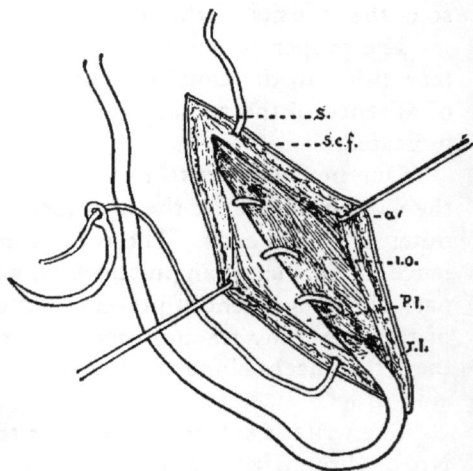

FIG. 5.—Deep tier of suture drawn home, obliterating inguinal canal. *S.*, skin; *s. c. f.*, subcutaneous fat; *a. e. o.*, apeneurosis of external oblique; *i. o.*, internal oblique; *P. l.*, Poupart's ligament.

at the lower end and outer side of the fascial wound (Figs. 4 and 5). A stitch is then taken, with still the same strand of catgut, piercing the internal pillar of external ring, round ligament, and external pillar. The excess of round ligament is now cut away just outside of the external ring, leaving the stump to plug the ring (Fig. 6).

After thus obliterating the inguinal canal and closing both internal and external rings, the same strand of catgut is continued upward as a running suture, uniting the lips of the incision in the external oblique aponeurosis and closing the anterior wall of the canal. At the upper end of the wound the two free ends of catgut emerging upon the aponeurosis of the external oblique are tied together, forming the only buried knot. This knot, if carefully and *tightly* tied after the manner depicted in the cut—a single turn in the first half and a double turn in the second half of the knot—can be depended upon not to slip. The skin is nicely approximated over all by a subcutaneous suture of ordinary catgut and the wound closed without drainage. Sterilized dressings applied over the wounds, and held in place by adhesive plaster and a double spica bandage, complete the operation.

The dressing is changed, for the sake of cleanliness, at the end of a week. This second dressing is removed a week later, and the patient allowed to sit up. No pessary or support of any kind is worn at any time after operation.

Interference with the function of the bladder as a result of the operation has not been observed in the writer's cases. The urine is generally drawn for two days, after which it is voided on the bed-pan.

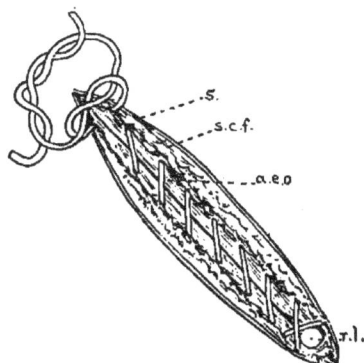

FIG. 6.—Superficial tier of buried suture of forty-day catgut closing incision through aponeurosis of external oblique, restoring anterior wall of canal. The excess of round ligament has been cut away just outside of external ring. The part protruding through ring together with pillars of external ring pierced by lowest loop of superficial suture. Loose knot at upper end shows proper way of tying buried catgut knot to prevent slipping. Skin and fat to be closed over all by a subcutaneous catgut suture.

Primary union has been the rule, except during a brief period of repeated deep suppuration from infected silkworm gut, which material we were using at the time in the form of buried sutures. In cases of primary union the scar becomes practically invisible after six months to a year.

In three of my last cases an attempt was made to complete the operation rapidly. The time required for these three operations, from the first incision to the completed closure of both wounds, was nineteen, twenty-one, and twenty minutes respectively.

TABLE OF AUTHOR'S OPERATIONS FOR SHORTENING THE ROUND LIGAMENTS.

No.	Name	Age	Condition.	No. of children.	Pathological conditions.	Date of operation.	Additional operations.	Result and remarks.
1	M. A.	20	Single.	..	Retroversion of uterus. Prolapse of left ovary.	29-xii-89	15-viii-93: Curettage of uterus.	25-vii-96: Uterus in normal anteversion. No prolapse of ovaries. Patient remains perfectly well.
2	L. R.	22	Single.	..	Retroversion of uterus.	31-xii-89	....	vi-93: Uterus in excellent anteversion.
3	B. S.	45	Married.	3	Complete prolapsus uteri et vaginæ.	1-ii-90	17-iii-90: Anterior colporrhaphy. Colpo-perineorrhaphy. vii-92: Nephropexy.	21-iv-90: Uterus well up in pelvis and anteverted. No prolapsus vaginæ.
4	L. S.	30	Married.	2	Retroversion of uterus. Movable right kidney.	3-ii-90		27-vii-96: Uterus in normal anteversion.
5	K. W.	43	Married.	2	Complete prolapsus uteri et vaginæ. Laceration and hypertrophy of cervix.	19-ii-90	At same sitting: Amputation of cervix. Colpo-perineorrhaphy.	19-iv-93: Uterus well up in pelvis and anteverted. Perfectly well with exception of moderate cystocele.
6	B. W.	29	Single.	..	Retroversion of anteflexed uterus.	24-ii-90	....	24-ii-92: Uterus in normal anteversion, the angle of flexion having disappeared.
7	M. P.	44	Married.	8	Complete prolapsus uteri et vaginæ. Laceration and hypertrophy of cervix.	5-iii-90	At same sitting: Amputation of cervix. Colpo-perineorrhaphy.	28-iv-96: Has remained perfectly cured of prolapsus. Uterus well up in pelvis and anteverted.
8	A. B.	41	Widow.	4	Prolapsus uteri, first degree, with retroversion. Laceration of cervix and Perinæum. Cystocele and rectocele.	24-iii-90	3-xii-89. Trachelorrhaphy. Anterior colporrhaphy. Colpo-perineorrhaphy.	7-xii-90: Prolapsus of vaginal walls cured. Uterus well up in pelvis and anteverted.
9	L. H.	57	Married.	3	Prolapsus and retroversion of uterus, both first degree. Moderate cystocele and rectocele.	9-v-90	....	28-vi-93: Uterus well up in pelvis and anteverted. No increase of cystocele and rectocele.
10	A. Le B.	37	Married.	5	Laceration of cervix. Retroversion of uterus. Movable right kidney.	19-v-90	At same sitting: Trachelorrhaphy. 12-vi-90: Nephropexy.	12-vii-90: Uterus in normal anteversion.

					Date	Operation	Result
11	J. M.	38	Married.	4	Laceration of cervix and perinæum. Rectocele. Retroversion of uterus. Hæmorrhoids.	21–v–90	At same sitting: Tra-chelorrhaphy. Col-po-perineorrhaphy.	Delivered, 11–ii–92, of full-grown child. Delivered, 12–i–94, of full-grown child. Uterus in normal anteversion after both deliveries. Normal labors. vii–96: Pregnant seven months; third pregnancy since operation.
12	M. S.	24	Married.	2	Chronic metritis. Laceration of cervix and perinæum. Retroversion of uterus.	28–v–90	At same sitting: Tra-chelorrhaphy. Perineorrhaphy.	28–vii–90: Uterus in normal anteversion.
13	S. K.	20	Single.	..	Retroversion of uterus. Endometritis.	12–vi–90	....	9–v–92: Normal delivery of living child at term, without complications. Development of right inguinal hernia during fifth month of this pregnancy, and of left inguinal hernia soon after delivery. Jan., 1896: Double inguinal hernia. Uterus in anteversion. July, 1896: Pregnant six months.
14	A. Q.	33	Married.	5	Retroversion of uterus. Laceration of cervix and perinæum.	19–ix–90	At same sitting: Tra-chelorrhaphy. 9–x–90: Perineorrhaphy.	Uterus in normal anteversion six months after operation.
15	M. D.	25	Married.	2	Retroversion of uterus. Chronic metritis.	6–xi–90	At same sitting: Amputation of cervix.	Uterus in normal anteversion on discharge one month after operation.
16	C. A.	44	Married.	10	Prolapsus and retroversion of uterus, first degree. Prolapsus vaginæ.	18–xi–90	At same sitting: Tra-chelorrhaphy. Anterior colporrhaphy. Colpo - perineorrhaphy.	Failure to find ligament on left side. Not attempted on right. Failure of operation to cure, the retroversion and prolapse returning promptly after patient left bed.
17	S. McM.	52	Widow.	1	Retroversion of uterus. Laceration of perinæum. Cystocele.	19–xii–90	20–xi–90: Anterior colporrhaphy. Col-po-perineorrhaphy.	Uterus well anteverted on leaving hospital one month after operation. No prolapsus of vaginal walls.
18	N. H.	23	Single.	..	Retroflexion of uterus. Chronic metritis.	19–xii–90	....	27–vi–93: Uterus in normal anteversion. Perfectly well ever since operation.
19	M. K.	32	Single.	..	Retroversion of uterus. Chronic endometritis.	7–iii–91	....	28–vi–93: Uterus in normal anteversion, limited in mobility. Patient not better than before operation owing to development of bilateral salpingo-oöphoritis since operation.

No.	Name.	Age.	Condition.	No. of children.	Pathological conditions.	Date of operation.	Additional operations.	Result and remarks.
20	A. M.	31	Married.	1	Subinvolution and retroflexion of uterus.	11-iv-91	Curettage of uterus.	1-vii-92: Delivered of a child at term, without complications. 3-viii-93: Uterus in normal anteversion. No pelvic symptoms. Patient died of pulmonary phthisis in 1894.
21	M. G.	17	Single.	..	Retroversion of uterus. Endometritis. Prolapse of ovaries.	13-v-91	Curettage of uterus.	22-vii-95: Uterus in anteversion. Left ovarian cystoma and movable right kidney have developed since operation. Patient no better than before operation.
22	P. K.	22	Single.	..	Retroversion of uterus.	20-v-91	....	On discharge, one month after operation uterus in normal anteversion. Not seen since.
23	A. K.	20	Single.	..	Retroversion of uterus. Endometritis. Catarrhal salpingitis.	17-vi-91	27-ii-93: Curettage of uterus.	Salpingitis and movable right kidney developed subsequently to shortening of round ligaments and nullified therapeutic results.
24	D. N.	25	Widow.	..	Retroversion of uterus. Laceration of cervix.	ζ-ix-91	At same sitting: Trachelorrhaphy.	27-v-93: Uterus in normal anteversion.
25	A. F.	31	Married.	..	Retroversio uteri fixati. Pelvi-peritonitis chronica.	17-x-91	12-x-91: Curettage of uterus and separation of adhesions in narcosis after Schultze.	Uterus in normal anteversion one month after operation. Not seen since. Patient writes that according to a statement of her physician, in July, 1895, her womb remains in anteversion.
26	J. M.	23	Single.	..	Retroversio of uterus. Prolapse of ovaries.	18-xi-91	Curettage of uterus at same sitting. 3-v-92: Salpingo-oöphorectomy for uterine fibroma, and curettage of uterus.	Fibromata uteri developed after operation. Uterus remained continuously in normal anteversion until last seen, 19-iii-95.
27	C. V.	29	Married.	2	Retroversion of uterus. Laceration of cervix. Endometritis. Catarrhal salpingitis.	30-xi-91	At same sitting: Curettage of uterus. Amputation of cervix.	Patient remained well for some six months after operation. Then acute prolapsus uteri, cured by tampons and supports. Later left chronic salpingo-oöphoritis nullified therapeutic results. 10-vii-96: Uterus in excellent anteversion. Descensus of first degree on straining.

28	M. W.	26	Married.	3	Retroversion of uterus. Laceration of cervix. Endometritis.	11-xii-91	At same sitting: Curettage of uterus. Amputation of cervix. 13-iii-96: Right nephropexy.	Felt well for several years after operation; then symptoms of movable right kidney. 17-vi-96: Uterus remains in normal anteversion.
29	M. S.	23	Single.	..	Retroversion of uterus. Prolapse of ovaries. Catarrhal salpingitis. Lues acquisita.	22-xii-91	Curettage of uterus.	Therapeutic results nullified by the progressive development of extensive syphilitic visceral lesions within a year after operation. The uterus remained in normal anteversion until her death, 19-xi-95.
30	C. R.	28	Married.	3	Subinvolution and retroversion of uterus.	15-i-92	At same sitting: Curettage of uterus. Amputation of cervix.	Uterus in normal anteversion six weeks after operation. Not seen since.
31	M. S.	25	Married.	1	Retroversion and subinvolution of uterus. Laceration of perinæum through sphincter.	19-i-92	At same sitting: Curettage of uterus. Amputation of cervix. Perineorrhaphy. Ventral fixation of uterus.	Right round ligament in final pull thereon drawn out of uterus as demonstrated at the immediately added ventral fixation. No operation upon left side. Ventral fixation holds uterus anteverted four and a half years after operation.
32	M. B.	23	Widow.	1	Retroversion of uterus, second degree. Prolapse of uterus, first degree. Chronic metritis. Mitral insufficiency and aortic stenosis.	22-i-92	At same sitting, without anæsthesia: Curettage of uterus. Amputation of cervix.	Shortening of round ligaments under local cocaine anæsthesia. 28-iv-92: Uterus in normal anteversion and at proper height in pelvis. Therapeutic results impaired by progressive left salpingo-oöphoritis.
33	K. S.	26	Married.	..	Retroversion of uterus. Endometritis. Prolapsed and adherent right ovary.	2-ii-92	Curettage of uterus.	Ether pneumonia after operation. 8-vii-96: Uterus in normal anteversion. Chronic metritis and adhesion of right ovary impair somewhat the therapeutic result.
34	C. H.	30	Married.	4	Retroversion of uterus. Laceration of cervix.	5-ii-92	At same sitting: Curettage of uterus. Amputation of cervix.	Uterus in normal anteversion. 3-v-94: Perfect anatomical and therapeutic results.

No.	Name.	Age.	Condition.	No. of children.	Pathological conditions.	Date of operation.	Additional operations.	Result and remarks.
35	M. F.	30	Married.	3	Retroversion of uterus. Laceration of cervix and perinæum.	22-iii-92	At same sitting: Curettage of uterus. Amputation of cervix. Perineorrhaphy. 3-vii-95: Curettage with drainage of uterus.	29-x-95: Uterus in normal anteversion. Felt perfectly well until three years after operation, when endometritis, salpingitis, and movable right kidney impaired therapeutic results.
36	A. H.	27	Married.	1	Retroversion of uterus. Laceration of cervix. Endometritis. Prolapse of ovaries.	19-iv-92	At same sitting: Curettage of uterus. Amputation of cervix.	26-vii-92: Uterus in normal anteversion.
37	A. B.	25	Single.	..	Retroversion of uterus. Endometritis. Catarrhal salpingitis. Movable right kidney.	10-v-92	At same sitting: Curettage of uterus. Nephropexy.	10-vii-92: Uterus in normal anteversion. 12-vii-93: Has remained perfectly well since operation. Pelvic examination not made.
38	P. K.	30	Married.	..	Retroversion of uterus. Endometritis.	10-v-92	Curettage of uterus.	Delivered of a full-grown boy at term in March, 1895. 23-vii-96: Uterus in normal anteversion. Patient enjoying perfect health.
39	M. K.	35	Married.	3	Adherent retroverted uterus. Laceration of cervix and perinæum. Chronic metritis. Cystoma of right ovary. Movable right kidney. Mitral insufficiency.	17-v-92	At same sitting: Curettage of uterus. Amputation of cervix. Ovariotomy. Ventral fixation of uterus. 18-xi-92: Nephropexy. Perineorrhaphy.	After isolating and drawing out both round ligaments, the right ligament, at the final pull, tore out of uterus as verified at the immediately following ventral fixation. Cause of the accident, a small, unrecognized, adherent right ovarian cystoma. 28-ii-94: Sudden death, while at work, from valvular disease of heart, on the eve of confinement at term. (Case reported in full, *Trans. New York Obst. Soc.*, Nov. 21, 1893, and April 17, 1894.)

40	A. M.	28	Married.	2	Retroversion of uterus. Laceration of cervix. Endometritis. Movable right kidney.	3–vi–92	At same sitting: Curettage of uterus. Amputation of cervix.	Uterus in normal anteversion one month after operation. Not seen since.
41	A. B.	24	Single.	..	Retroversion of uterus. Endometritis.	17–vi–92	Curettage of uterus.	Passed through a severe attack of typhus abdominalis beginning two weeks after operation. 22–vi–93: Uterus in normal anteversion.
42	A. H.	17	Single.	..	Retroversion of uterus. Endometritis.	22–vii–92	Curettage of uterus. 11–vii–93: Right nephropexy. 15–iv–95: Right femoral herniotomy. 6–xii–95: Cœliotomy for acute gangrenous appendicitis. Right salpingo-oöphorectomy. 7–i–96: Left nephropexy.	Typhoid fever during convalescence. Movable right and left kidneys followed the typhoid fever. Acute strangulated right femoral hernia, April, 1893. Acute appendicitis with perforation, periappendicular abscess and serious involvement of right tube and ovary, Dec., 1895. 22–vii–96: Uterus in normal anteversion. Patient perfectly well.
43	M. S.	35	Married.	4	Retroversion of uterus. Laceration of cervix and perinæum. Rectocele.	28–x–92	At same sitting: Curettage of uterus. Trachelorrhaphy. Colpo-perineorrhaphy.	28–vi–93: Uterus in anteversion.
44	L. S.	32	Married.	1	Adherent retroverted uterus. Laceration of cervix. Endometritis.	8–xi–92	At same sitting: Separation of uterine adhesions in narcosis after Schultze. Curettage of uterus. Trachelorrhaphy.	Uterus in normal anteversion three months after operation. Not seen since.
45	R. S.	36	Married.	6	Retroversion of uterus. Laceration of cervix. Endometritis.	9–xi–92	At same sitting: Curettage of uterus. Amputation of cervix. 16–xii–92: Hystero-salpingo-oöphorectomy by combined perineotomy and cœliotomy for sarcoma of left broad ligament.	Uterus remained in normal anteversion until removed together with the tubes, ovaries, and entire left broad ligament by combined perineotomy and cœliotomy for sarcoma of left broad ligament, 16–xii–92. (Case reported in detail, *Trans. New York Obst. Soc.*, March 21, 1893, and March 17, 1896.)

No.	Name	Age	Condition.	No. of children.	Pathological conditions.	Date of operation.	Additional operations.	Result and remarks.
46	A. W.	25	Married.	3	Retroversio uteri. Laceration of cervix and perinæum. Left inguinal hernia.	25-x-92	At same sitting: Curettage of uterus. Amputation of cervix. Anterior colporrhaphy. Perineorrhaphy. Radical inguinal herniotomy.	21-vi-93: Miscarriage at end of third month. Thereafter uterus in normal anteversion. 20-ii-94: Pregnant seven months.
47	A. H.	19	Single.	..	Retroversion of uterus. Endometritis.	29-x-92	Curettage of uterus.	9-xii-93: Uterus in normal anteversion. Endometritis for past two months.
48	A. W.	28	Married.	1	Retroversion of uterus. Laceration of cervix. Endometritis.	13-xii-92	At same sitting: Curettage of uterus. Amputation of cervix.	2-vii-96: Uterus in normal anteversion. Perfectly well for three years after operation, when mobility of both kidneys developed, from the symptoms of which she now suffers.
49	A. B.	25	Married.	2	Retroversion of uterus. Laceration of cervix and perinæum. Endometritis. Cystocele.	3-i-93	At same sitting: Curettage of uterus. Anterior colporrhaphy. Perineorrhaphy. Ventral fixation of uterus.	Right round ligament pulled out of uterus after both ligaments had been found and drawn out. Ventral fixation immediately added. Three pregnancies since operation. Delivery at term, without complications. 11-xii-93. Abortus, 10-vii-95. Now, vii-96, pregnant eight months. (Case detailed in *Trans. New York Obst. Soc.,* April 17, 1894.)
50	J. D.	20	Married.	1	Retroversion and subinvolution of uterus. Laceration of cervix and perinæum. Endometritis. Urethral polypi.	17-1-93	At same sitting: Curettage of uterus. Amputation of cervix. Perineorrhaphy. Excision of urethral polypi.	One month after operation uterus in normal anteversion. Not seen since.
51	S. H.	22	Single.	1	Retroversion of uterus. Endometritis.	10-ii-93	Curettage of uterus.	Uterus in normal anteversion one month after operation. Not seen since.

52	A. V.	21	Married.	1	Retroversion of uterus. Laceration of cervix. Endometritis.	10-iii-93	At same sitting: Curettage of uterus. Trachelorrhaphy.	Uterus in normal anteversion when last seen, three months after operation.
53	L. S.	33	Widow.	5	Retroversion of uterus. Laceration of cervix and perinæum. Endometritis.	14-iii-93	At same sitting: Curettage of uterus. Trachelorrhaphy. Perineorrhaphy.	12-vii-93: Feels perfectly well. Uterus in normal anteversion.
54	M. B.	22	Single.	..	Retroversion of uterus. Endometritis.	21-iv-93	Curettage of uterus.	Uterus in normal anteversion six weeks after operation. Not seen since.
55	T. W.	37	Widow.	1	Retroversion of uterus. Endometritis.	25-iv-93	Curettage of uterus.	Uterus in normal anteversion one month after operation. Not seen since.
56	M. H.	22	Single.	..	Retroversion of uterus. Endometritis.	28-iv-93	Curettage of uterus.	Right round ligament, after leaving the internal ring, ran upward and outward to be inserted into outer half of Poupart's ligament. 24-iv-94: Uterus in normal anteversion.
57	M. K.	34	Married.	5	Retroflexion of uterus. Laceration of cervix and perinæum. Mitral insufficiency.	23-v-93	21-iv-93: Curettage of uterus. Amputation of cervix. Perineorrhaphy.	Therapeutic result excellent until two weeks ago, since when acute oöphoritis sinistra. Uterus in normal anteversion ten weeks after operation, when last seen.
58	D. M.	17	Single.	..	Retroversion of uterus. Endometritis.	13-vi-93	Curettage of uterus.	Felt perfectly well until fifteen months after operation. Then acute articular rheumatism and endocarditis. 14-x-94: Uterus in normal anteversion.
59	M. M.	24	Married.	1	Retroversion of uterus. Laceration of cervix. Endometritis.	20-vi-93	At same sitting: Curettage of uterus. Trachelorrhaphy.	10-vii-96: Uterus in normal anteversion. Therapeutic result good until recently, when impaired by development of movable right kidney and recurrence of endometritis.
60	B. W.	21	Single.	..	Retroversion of uterus. Endometritis.	27-vi-93	Curettage of uterus.	9-vii-96: Patient has "felt splendid" ever since operation. Uterus in normal anteversion.

No.	Name.	Age.	Condition.	No. of children.	Pathological conditions.	Date of operation.	Additional operations.	Result and remarks.
61	J. B.	25	Single.	..	Retroversion of uterus. Endometritis. Bilateral inguinal hernia.	18–vii–93	At same sitting: Curettage of uterus. Double radical inguinal herniotomy.	Patient irresponsible mentally; repeatedly removed her wound dressing. Suppuration and sloughing in both wounds. 26–iv–94: Uterus in normal anteversion. Left hernia recurred six months after operation. Right hernia remains cured.
62	M. D.	25	Married.	3	Retroversion of uterus. Laceration of cervix and perinaeum. Endometritis. Cystocele.	15–viii–93	At same sitting: Curettage of uterus. Amputation of cervix. Anterior colporrhaphy. Perineorrhaphy.	25–vii–96: Uterus in normal anteversion. Patient well except for symptoms due to mobility of both kidneys existing before operation. Bilateral nephropexy advised.
63	M. H.	23	Single.	..	Retroversion of uterus. Endometritis.	7–xi–93	Curettage of uterus.	30–vi–96: Uterus in normal anteversion. Therapeutic results excellent until within three months past, since which time development of movable right kidney.
64	C. L.	27	Married.	..	Retroversion of uterus. Endometritis.	23–xi–93	Curettage of uterus.	1–vii–96: Anatomical and therapeutical results excellent. Uterus in normal anteversion.
65	A. E.	31	Married.	1	Retroversion of uterus. Laceration of cervix. Endometritis.	9–i–94	At same sitting: Curettage of uterus. Trachelorrhaphy.	Patient widowed one month after operation. 10–vii–96: Anatomical and therapeutic results excellent. Uterus in normal anteversion.
66	L. H.	21	Married.	1	Retroversion of uterus. Endometritis.	17–i–94	Curettage of uterus.	Uterus in normal anteversion one month after operation. Not seen since.
67	N. T.	29	Single.	..	Retroversion of uterus. Endometritis. Prolapse of ovaries. Oöphoritis dextra.	22–i–94	Curettage of uterus.	Delivery at term, without difficulty, 2–xi–95. 27–vii–96: Uterus in normal anteversion. Both ovaries and all pelvic contents normal. Patient perfectly well.
68	M. A.	28	Widow.	1	Retroversion of uterus. Laceration of cervix and perinaeum. Endometritis. Rectocele.	23–i–94	At same sitting: Curettage of uterus. Amputation of cervix. Perineorrhaphy.	Uterus in normal anteversion seven weeks after operation, when patient was last seen.

69	A. M.	23	Single.	..	Anteflexion of uterus. Chronic endometritis of cervix. Movable right kidney.	9–iii–94	At same sitting: Curettage of uterus. Amputation of cervix. 15–vi–94: Nephropexy.	28–vi–96: Anteflexion has entirely disappeared, the uterus lying in normal anteversion. Patient perfectly well.
70	E. S.	26	Married.	..	Retroversio uteri. Endometritis. Catarrhal salpingitis. Tuberculosis of appendix vermiformis.	30–iii–94	At same sitting: Curettage of uterus. Appendectomy.	Ligaments sloughed in both wounds. Six weeks after operation position of uterus unsatisfactory, not sufficiently well anteverted, rather vertical in body. Very hysterical before operation. Insanity (hereditary taint) followed operation. Sudden death from fatty heart, 22–viii–94, nearly five months after operation. Autopsy showed the uterus again fully retroverted.
71	A. B.	29	Married.	4	Retroversion of uterus. Laceration of cervix. Chronic metritis. Movable right kidney. Right inguinal hernia. Right femoral hernia.	4–v–94	At same sitting: Curettage of uterus. Amputation of cervix. Radical femoral herniotomy. Radical inguinal herniotomy.	23–i–96: Patient pregnant two months. Right inguinal hernia has returned. Femoral hernia remains cured. Uterus in normal anteversion. 1–vii–96: Patient eight months pregnant without inconvenience.
72	L. M.	33	Married.	3	Retroversion of uterus. Laceration of cervix. Prolapse of ovaries. Endometritis. Salpingitis.	8–v–94	20–vii–93: Curettage of uterus. Amputation of cervix.	Therapeutic success nil. Dr. H. J. Boldt subsequently performed vaginal bilateral salpingo-oöphorectomy, after which patient reports herself well. 20–vii–94: Dr. Boldt writes me that the uterus remains anteverted and well up in pelvis.
73	S. R.	20	Single.	..	Imperfect development of internal genitalia. Retroversion of uterus. Endometritis.	12–vi–94	12–1–92: Curettage of cervix. Bilateral discission of cervix. 12–vi–94: Curettage of uterus. Amputation of cervix.	Uterus in normal anteversion one month after operation, when last seen.

No.	Name.	Age.	Condition.	No. of children.	Pathological conditions.	Date of operation.	Additional operations.	Result and remarks.
74	A. H.	26	Single.	1	Movable right kidney. Retroversion of uterus. Laceration of cervix. Endometritis. Right and left inguinal hernia.	13-vi-94	17-i-94: Nephropexy. 13-vii-94: Curettage of uterus. Trachelorrhaphy. Bilateral radical inguinal herniotomy.	14-iv 96: Uterus in normal anteversion. No return of either inguinal hernia. Patient perfectly well.
75	J. McP.	21	Single.	1	Retroversion of uterus. Laceration of uterus. Chronic metritis. Movable right and left kidneys. Chronic appendicitis.	2-x-94	At same sitting: Curettage of uterus. Amputation of cervix. 21-xii-94: Bilateral nephropexy. 15-ii-95: Inversion of appendix.	28-vi-96: Uterus in normal anteversion, still enlarged about 50 per cent. from persistence of chronic metritis.
76	S. M.	31	Married.	5	Retroversion of uterus. Laceration of cervix. Endometritis.	11-x-94	25-i-94: Curettage of uterus. Amputation of cervix.	28-i-95: Uterus in normal anteversion. Malaria and the development of movable right kidney since operation impair the therapeutic result.
77	E. F.	23	Single.	..	Movable right kidney. Retroversion of uterus. Endometritis. Catarrhal salpingitis. Appendicitis. Mitral insufficiency.	16-x-94	3-viii-94: Nephropexy. At same sitting: Curettage of uterus. Appendectomy.	Uterus in normal anteversion one month after operation, since which patient not seen.
78	B. C.	25	Single.	..	Retroversion of uterus. Endometritis. Left salpingo-oöphoritis.	18-x-94	Curettage of uterus.	27-vi-96: Uterus in normal anteversion. No trace of former salpingo-oöphoritis. Patient perfectly well.
79.	M. McC.	27	Married.	4	Laceration of cervix and perinæum. Endometritis. Retroversion of uterus.	19-x-94	3-viii-94: Curettage of uterus. Trachelorrhaphy. Perineorrhaphy.	Left ligament tore on final pull, one centimetre from uterine insertion. Ventral fixation immediately added. Right groin not opened. 23-vi-96: Two pregnancies since operation. First terminated in abortion in fourth month. Now pregnant in eighth month.

					Diagnosis.	Date.	Operation.	Result.
80	R. S.	18	Single.	..	Movable right kidney. Chronic appendicitis. Endometritis. Retroversion of uterus.	23–x–94	17–iv–94: Nephro-pexy. 23–vi–94: Appendectomy. At same sitting, Dilatation of sphincter ani.	Uterus in normal anteversion and patient perfectly well when last seen, five months after operation.
81	M. D.	18	Single.	..	Retroversion of uterus. Urethral polypi. Adherent prepuce of clitoris. Hæmorrhoids.	2–xi–94	At same sitting: Curettage of uterus. Ablation of urethral polypi. Dilatation of sphincter ani. 27–xi–94: Slitting of prepuce and separation of preputial adhesions. Excision of hæmorrhoids.	21–vii–96: Patient made a new woman by the operations. Subsequently slight relapse due to development of movable right kidney and recurrence of endometritis. Uterus in normal anteversion.
82	B. H.	20	Single.	..	Retroflexion of uterus. Endometritis. Chronic appendicitis. Incipient pneumo-phthisis.	16–xi–94	At same sitting: Curettage of uterus. Appendectomy.	25–vii–96: Patient remains well. Uterus in normal anteversion.
83	A. M.	25	Single.	..	Adherent retroverted uterus. Endometritis.	25–xi–94	At same sitting: Curettage of uterus. Cœliotomy for liberation of uterine adhesions. Attempted ventral fixation of uterus. 25–iii–95: Vaginal fixation of uterus.	Ventral fixation of uterus, after liberation of adhesions, was the operation proposed. It failed on account of inability to bring the fundus against anterior abdominal wall, due to thickening and inelasticity of parametria. Shortening of round ligaments immediately added. Severe acute croupous pneumonia and pneumococcus infection of three wounds. Sloughing of round ligaments and immediate anatomical failure. Four months later vaginal fixation of uterus with resultant imperfect anatomical but perfect therapeutic success. (Case detailed in *N. Y. Med. Monatsschrift*, July, 1896, p. 285.)

No.	Name.	Age.	Condition.	No. of children.	Pathological conditions.	Date of operation.	Additional operations.	Result and remarks.
84	B. K.	26	Married.	1	Retroversion of uterus. Laceration of cervix and perinæum. Endometritis.	18-xii-94	At same sitting: Curettage of uterus. Trachelorrhaphy. Perineorrhaphy. Ventral fixation of uterus.	Left round ligament torn out of uterus at final pull. Ventral fixation. 10-vii-96: Patient has been perfectly well since operation. Fundus attached to anterior abdominal wall.
85	A. W.	24	Single.	..	Retroversion of uterus. Endometritis.	8-xi-95	Curettage of uterus.	Uterus in normal anteversion when last seen, one month after operation.
86	L. McK.	20	Single.	..	Retroversion of uterus. Endometritis. Left salpingo-oöphoritis, with adhesions.	12-iii-95	At same sitting: Curettage of uterus. Posterior colpotomy to liberate adherent left tube and ovary.	Uterus in normal anteversion, six weeks after operation, when last seen.
87	L. H.	22	Single.	..	Retroversio uteri. Endometritis.	26-iii-95	Curettage of uterus.	Uterus in normal anteversion two months after operation, when last seen.
88	M. T.	39	Single.	..	Retroversio uteri. Endometritis.	29-iii-95	Curettage of uterus.	Right ligament tore just within internal ring. Small median incision made, and by passing finger through it into abdomen the uterus was lifted forward and the uterine half of round ligament was pushed out through right internal ring and sutured in the canal in the usual way.
89	L. K.	28	Married.	3	Retroversion of uterus. Laceration of cervix. Chronic metritis.	2-iv-95	At same sitting: Curettage of uterus. Amputation of cervix.	Uterus in normal anteversion when last seen, one month after operation. 7-vii-96: Uterus in normal anteversion, still very heavy. Therapeutic result marred by subsequent development of movable right kidney and chronic appendicitis
90	J. K.	25	Single.	..	Retroversion of uterus. Endometritis. Catarrhal salpingitis. Movable right kidney. Proctitis catarrhalis.	18-iv-95	Curettage of uterus. 24-ii-96: Nephropexy. Dilatation of sphincter ani.	June, 1896: Uterus in normal anteversion. Therapeutic results excellent.

No.		Age			Diagnosis	Date	Operation	Remarks
91	A. K.	22	Married.	1	Retroversion of uterus. Laceration of cervix. Endometritis.	30-iv-95	At same sitting: Curettage of uterus. Amputation of cervix.	Uterus in normal anteversion when last seen, six weeks after operation.
92	E. H.	24	Married.	1	Retroversion of uterus. Laceration of cervix. Endometritis. Catarrhal salpingitis.	7-v-95	At same sitting: Curettage of uterus. Amputation of cervix.	Neither ligament being found in canal, an incision was made in the median line and the ligaments traced from either cornu uteri outward. After passing through the internal inguinal ring each ligament ran upward and outward between the internal oblique and the transversalis to be inserted into the outer half of Poupart's ligament and the transversalis fascia as far outward as the anterior superior spine of the ilium. Both ligaments cut away above, brought down into canal and fastened in usual way. Uterus in normal anteversion when last seen, one month after operation.
93	S. G.	29	Single.	..	Retroversion of uterus. Endometritis.	21-vi-95	Curettage of uterus.	15-ii-96: Uterus in normal anteversion. Patient perfectly well.
94	M. C.	28	Single.	..	Retroversion of uterus. Laceration of cervix. Endometritis.	25-x-95	At same sitting: Curettage of uterus. Amputation of cervix.	Uterus in normal anteversion, one month after operation, when last seen. July, 1896, writes that she is well.
95	C. C.	33	Married.	1	Retroversion of uterus. Endometritis. Catarrhal salpingitis. Movable right kidney	6-xi-95	17-x-95: Curettage of uterus. Nephropexy.	Uterus in normal anteversion at last examination two months after operation. July, 1896, writes that she remains perfectly well.
96	M. W.	23	Married.	1	Retroversion of uterus. Laceration of cervix. Endometritis.	22-xi-95	At same sitting: Curettage of uterus. Amputation of cervix.	7-v-96: Uterus in normal anteversion. Therapeutic result marred by development of movable right and left kidneys since operation.
97	L. K.	32	Married.	1	Retroversion of uterus. Laceration of cervix. Endometritis. Movable right and left kidneys.	10-xii-95	At same sitting: Curettage of uterus. Amputation of cervix.	Twin pregnancy immediately after returning home. Abortion at end of third month, 3-iv-96. 1-vii-96: Uterus in normal anteversion. Patient well with exception of symptoms due to movable kidneys.

No.	Name.	Age.	Condition.	No. of Children.	Pathological conditions.	Date of operation.	Additional operations.	Result and remarks.
98	M. K.	31	Single.	..	Retroversion of uterus. Chronic metritis.	17-xii-95	At same sitting: Curettage of uterus. Amputation of cervix.	Left round ligament normal in every way. Right round ligament not found in canal. Through a small median incision of the abdomen the right ligament was traced from the cornu of uterus outward, where, after passing through the internal inguinal ring, it ran upward and outward to be inserted into outer half of Poupart's ligament. It was cut away from its abnormal attachments, brought down and sutured in the inguinal canal in the usual way. Anatomical result perfect after one month, when patient was last seen.
99	A. C.	30	Married.	4	Retroversion of uterus. Chronic metritis. Bilateral salpingo-oöphoritis, with adhesions.	24-xii-95	At same sitting: Curettage of uterus. Amputation of cervix. 17-i-96: Right salpingectomy. Left salpingostomy. Ventral fixation of uterus.	Primary union of both wounds. Patient sat up in bed on eleventh day. Two weeks after operation uterus was found again retroverted. 17-i-96. Cœliotomy. Chronic salpingo-oöphoritis with universal adhesions of appendages posteriorly in pelvis. Uterus not adherent. Tubes and ovaries not enlarged except outer ends of tubes which were occluded. Left ligament remained securely fastened in canal. Right round ligament retracted within abdomen; stump thereof contracted to four centimetres in length. Failure due to firm adhesions of tubes and ovaries. After liberation of these, right salpingectomy, left salpingostomy and ventral fixation of uterus; neither ovary removed. 30-vi-96: Patient remains well. Fundus forward, attached to abdomina wall.

100	M. M.	38	Married.	2	Retroversio uteri. Chronic metritis. Dementia.	31-xii-95	At same sitting: Curettage of uterus. Amputation of cervix.	Patient neither before nor after operation, while under our care, passed water voluntarily, requiring the constant use of catheter. Notwithstanding several severe over-distentions of bladder, uterus in normal anteversion on discharge a month after operation, since which time not seen.
101	M. McC.	19	Single.	..	Retroversion of uterus. Endometritis.	6-i-96	Curettage of uterus.	2-vii-96: Uterus in normal anteversion. Patient well.
102	E. K.	45	Single.	1	Retroversion of uterus. Chronic metritis and perimetritis.	17-i-96	At same sitting: Curettage of uterus. Posterior colpotomy for separation of tubal and ovarian adhesions.	Ovaries, tubes, and uterus adherent; otherwise normal. Adhesions separated through an incision of the posterior *cul-de-sac*, which was then closed. 8-vii-96: Uterus in normal anteversion. Tubes and ovaries normal. Patient well.
103	M. C.	20	Single.	1	Retroversion of uterus. Laceration of cervix, vagina, and perinæum.	7-ii-96	At same sitting: Curettage of uterus. Trachelorrhaphy. Median coeliotomy.	Difficulty in identifying ligaments in canal. Small median abdominal incision, through which ligaments were traced from uterine cornua outward to canals and there identified. Uterus in normal anteversion one month after operation, when last seen.
104	E. S.	31	Single.	1	Retroversion of uterus. Chronic metritis. Chronic parenchymatous nephritis.	11-ii-96	......	14-vii-96: Uterus in normal anteversion. Patient relieved of pelvic symptoms. Nephritis persists.
105	J. McG.	18	Single.	..	Anteflexion of uterus. Endometritis.	14-ii-96	Curettage of uterus.	One month after operation uterus in normal anteversion, the angle of flexion having disappeared. Not seen since.
106	M. B.	22	Single.	..	Retroflexion of uterus. Endometritis. Small cystic degeneration of right ovary.	18-ii-96	At same sitting: Median cœliotomy. Puncture of ovarian cysts.	Right ligament would not run out on traction. Small median abdominal incision made to ascertain cause, which was found to consist in abnormal shortening of right infundibulo-pelvic ligament. This ligament stretched, few small cysts of right ovary punctured and the round ligaments shortened in usual way. 14-vii-96: Uterus in normal anteversion. Patient well in every way.

No.	Name.	Age.	Condition.	No. of children.	Pathological conditions.	Date of operation.	Additional operations.	Result and remarks.
107	A. L.	24	Single.	..	Retroversio uteri. Endometritis.	3-iii-96	Curettage of uterus.	Uterus in normal anteversion one month after operation. Not seen since.
108	F. K.	33	Married.	7	Retroversion of uterus. Laceration of cervix. Endometritis.	10-iii-96	At same sitting: Curettage of uterus. Amputation of cervix.	30-vi-96: Uterus in normal anteversion. Patient perfectly well.
109	M. J.	36	Married.	2	Retroversion of uterus. Laceration of cervix. Endometritis.	10-iv-96	Curettage of uterus.	Uterus in normal anteversion ten weeks after operation, when last seen.
110	M. McK.	32	Widow.	3	Retroversion of uterus. Laceration of cervix. Endometritis.	14-iv-96	At same sitting: Curettage of uterus. Amputation of cervix.	Uterus in normal anteversion and patient perfectly well ten weeks after operation, when last seen.
111	L. P.	42	Married.	1	Retroversion of uterus. Chronic metritis. Right Bartholinitis.	25-iv-96	Curettage of uterus.	Two months after operation, when last seen, uterus in normal anteversion and patient well.
112	E. D.	30	Married.	3	Retroversion of uterus. Laceration of cervix. Endometritis. Mitral and aortic insufficiency.	28-iv-96	At same sitting: Curettage of uterus. Amputation of cervix.	Two months after operation, uterus in normal anteversion. Shortening of round ligaments performed under infiltration anæsthesia (Schleich).
113	A. A.	22	Single.	1	Retroversion of uterus. Laceration of cervix. Chronic metritis.	8-v-96	At same sitting: Curettage of uterus. Amputation of cervix.	Uterus in normal anteversion and patient well when last seen, six weeks after operation.
114	A. J.	23	Married.	..	Retroversion of uterus. Endometritis.	15-v-96	Curettage of uterus.	Uterus in normal anteversion six weeks after operation.
115	L. P.	23	Married.	1	Retroversion of uterus. Laceration of cervix. Chronic metritis.	15-v-96	At same sitting: Curettage of uterus. Amputation of cervix.	Uterus in normal anteversion when last seen, six weeks after operation.

RESULTS AND ANALYSIS OF AUTHOR'S OPERATIONS FOR SHORTEN-
ING THE ROUND LIGAMENTS.

I have attempted the operation of shortening the round ligaments upon one hundred and twenty-three women all told. My first
five cases were operated upon in 1889, and were patterned upon
the operation as performed by Alexander. The records of these
cases have been lost. The histories of two further patients, operated
upon in other cities for the purpose of demonstrating the operation,
were never obtained. The records of these seven patients and that
of the only patient who died—eight cases in all—are not included
in the appended table of operations.

Mortality.—Of the one hundred and twenty-three patients operated upon, one died within a week after operation of acute septic
peritonitis, which, until the autopsy proved the contrary, was attributed to the operation. The necropsy showed acute gangrenous
appendicitis, with perforation and purulent general peritonitis, the
death being in no wise attributable to the operation. The mortality
may therefore be fairly stated as *nil.* This is the more remarkable
when the large number of additional operations, many of them of
a serious character, performed upon the same patients, and the
formidable list of diseases complicating convalescence, are taken
into account.

Methods of Operation.—Of the one hundred and fifteen cases
tabulated, Nos. 1 to 74 were operated upon after the writer's original method (123), except that the buried silkworm-gut suture was
used in the last thirty-eight. In Nos. 75 to 101 the method of
Alquié-Kellogg was followed. Nos. 102 to 115, finally, were operated upon with my present improved technics.

Age.—The youngest patient was seventeen, the oldest fifty-seven
years of age. The average age of the one hundred and fifteen
women was twenty-eight years. Fifty-nine patients were married,
forty-eight single, eight widows.

Indication for Operation.—

Retroversion and excessive mobility of the uterus, un-
    complicated, or complicated only by changes lim-
    ited to the uterus itself, as endometritis, chronic
    metritis, laceration of cervix, subinvolution...... 87
Retroversion, with adhesions of the uterus.......... 4

This list of indications for shortening the round ligaments is believed to sufficiently explain itself, except perhaps as regards the use of the term " prolapse of ovaries." The ovaries are practically found more or less prolapsed in every case of movable retroversion of the uterus. The case has been classified as prolapse of the ovaries, instead of under retroversion of the uterus, whenever the two ovaries prolapsed completely, so as to lie side by side, in contact with each other, in the deepest part of Douglas' sac, *beneath* the retroverted fundus uteri. Whenever the fundus uteri occupied the lowest part of Douglas' sac, with an ovary on either side, the case was classed under the heading retroversion of the uterus.

The relative rarity of cases of retroflexion, as compared with those of retroversion, is due to the fact that those cases only were classed as retroflexion in which the retrodeviated fundus was at least as low in the pelvis as the cervix, and in which, at the same time, the angle of flexion was so acute that the posterior surfaces of corpus and cervix were practically in juxtaposition.

In the cases of retrodeviations of the uterus with adhesions of that organ or of its adnexa, or of both, the adhesions, except in two cases where existing adhesions of the right ovary were not recognized, were first separated either by bimanual divulsion after Schultze, by posterior colpotomy, by median cœliotomy, or by incision of the peritonæum at the internal inguinal ring.

Additional Operations.—In addition to shortening the round ligaments, or attempting to shorten them, the writer performed upon these one hundred and fifteen patients, at the same or additional sittings, the following operations:

Slitting of præputium clitoridis and separation of preputial adhesions...................................... 1

Excision of urethral polypus...................... 2

Stretching of sphincter ani........................ 3

Excision of hæmorrhoids........................... 1

Perineorrhaphy 14

Colpo-perineorrhaphy 8

Anterior colporrhaphy............................. 7

Bilateral discission of cervix...................... 1

Trachelorrhaphy 17

Amputation of cervix.............................. 40

Curettage of uterus................................ 96

Posterior colpotomy............................... 2

Vaginal fixation of uterus.......................... 1

Separation of adhesions in narcosis (Schultze)........ 2

Cœliotomy for separation of adhesions.............. 3

Ventral fixation of uterus.......................... 5

Left salpingostomy................................ 1

Right salpingectomy............................... 1

Salpingo-oöphorectomy for fibroma uteri............ 1

Cœliotomy and puncture of ovarian small cysts....... 1

Ovariotomy 1

Hystero-salpingo-oöphorectomy, by combined perineotomy and cœliotomy, for sarcoma of left broad ligament 1

Inversion of vermiform appendix................... 1

Appendectomy for chronic appendicitis.............. 3

Appendectomy for tuberculosis of appendix......... 1

Cœliotomy for acute gangrenous appendicitis........ 1

Radical left inguinal herniotomy.................... 1

Radical right inguinal herniotomy.................. 1

Bilateral radical inguinal herniotomy............... 2

Right femoral herniotomy......................... 2

Nephropexy, for movable kidney................... 11

Bilateral nephropexy.............................. 2

Including the shortening of the round ligaments, three hundred and forty-nine operations were performed upon these one hundred and fifteen patients. All of the patients made good recoveries from all the operations performed upon them.

Technical Difficulties encountered in the Operation of shortening the Round Ligaments.—The writer has noted, at the time of their occurrence, all the various technical difficulties encountered in these one hundred and fifteen operations for shortening the round ligaments. His experience in this direction he believes to have been unusually large and varied, and a *résumé* of these difficulties and the way in which they were met may possibly prove of service to younger operators.

First, as regards finding the ligaments. In one instance only (Case XVI) was the round ligament not found, search for it being made on the left side only. The writer feels convinced that the ligament was present, and that the failure to find it was entirely chargeable to his lack of resources and development as an operator upon the round ligaments.

In a second instance (Case CIII), after failure to find in the inguinal canal the round ligament there present, the abdomen was opened in the median line by a small incision, through which the round ligaments were traced from the cornu uteri to the internal ring and into the inguinal canal, where they were then readily identified.

In searching for the round ligament the right deep epigastric artery was divided and doubly ligated in Case XXII. In Case LVI, in searching for a round ligament not present in the canal, the right internal iliac vein was drawn out by the hook, recognized, and dropped without injuring the vessel. In Case XCVI, in fishing for the ligament, the right internal iliac artery was hooked up, recognized as such, and dropped without injury or bad result.

The abnormal course and insertion of the right round ligament in Cases LVI and XCVIII, and of both round ligaments in Case XCII, necessitating their absence from the canal, have already been described. In similar cases, upon not finding the round ligament in the canal, I would propose, as the best plan to follow, to open the peritonæum at the internal ring. The round ligament would then either be recognized running between the folds of the broad ligament or could be readily traced outward from the cornu uteri.

The round ligament may tear, either in the canal or between the internal ring and the cornu uteri, in the attempt to isolate it or in drawing it out of the abdomen.

Three times the round ligament (the left in Case L, and the right in Cases LXX and CVIII) is recorded as being torn within the canal, the uterine end retracting through the internal ring into the abdomen. In one case the torn uterine end was easily recovered just within the internal ring. In a second case the ligament was found, and its torn end pushed out through the internal ring by the aid of a small incision in the median line of the abdomen. In a third instance, a tear of the *left* round ligament, the peritonæum was incised at the *right* internal inguinal ring, and by means of a finger passed through the latter into the abdomen the uterine end of the torn ligament was pushed through the left internal ring into the canal. In all three cases the recovered ligament was secured in the canal in the usual way.

No less than six times the round ligament (the right in Cases XXXI, XXXIX, XLIX, LXXXVIII, and the left in Cases LXXIX, LXXXIV) tore within the abdomen in the attempt to draw it out. In four of these cases the ligament was pulled clean out of the uterus, leaving a depression or hole at the cornu to mark the site of its former insertion. In the fifth case the ligament parted at a distance of one centimetre from the uterine cornu. In each of these five cases median cœliotomy and ventral fixation of the uterus were performed immediately on the occurrence of the accident. In the sixth case (Case LXXXVIII) the round ligament tore just within the internal inguinal ring, leaving the uterine end long enough for purposes of shortening. This uterine end was found through a small median incision, pushed out through the internal ring into the canal, and there secured in the usual way.

In one of the six cases an unrecognized, adherent, small ovarian cystoma was the cause of the tear. The other five must be explained either by diminished strength of the round ligament, due to fatty degeneration or other causes, or to undue force exercised by the operator. The methods of meeting the accident of tear of the round ligament have already been considered.

In developing or drawing out the round ligament its central fibers, or core, as it were, have sometimes been drawn out instead of the entire ligament, the outer or peripheral fibers being held by their attachments in the canal. In such cases the cone of peritonæum accompanying the round ligament should be incised, if necessary, to secure the entire ligament.

The peritonæum incasing the round ligament can generally be

stripped back with great facility. Occasionally, however, the peritoneal investment has been found so firmly attached as to resist blunt dissection. No hesitation need be felt by an operator sure of his asepsis in using the scissors or knife, even though the peritonæum should be opened. In one instance (Case XIII), where the peritonæum was opened, the uterine end of the Fallopian tube was drawn out of the internal ring with the round ligament. The two structures were bluntly separated, the tube replaced within the abdomen, the ligament shortened and anchored in the usual way.

On three or four occasions the writer has had the opportunity at subsequent cœliotomies of observing the condition of the shortened round ligaments. They have always been encountered as short cords running in an almost straight line from cornu uteri to internal inguinal ring, and projecting on the anterior face of the broad ligaments in such a manner as to form a shallow peritoneal depression or pouch between the bladder anteriorly, the fundus uteri posteriorly, and the projecting round ligaments on either side.

Behavior of Operation Wounds.—Ninety-three per cent. of the wounds healed by primary union; in seven per cent. wound infection and more or less extensive suppuration are noted. In one case of acute croupous pneumonia immediately following operation (Case LXXXIII), infection of both inguinal wounds, as well as of a median abdominal incision made at the same time, followed. The three abscesses were opened on the tenth day after operation, and the pus from each of them furnished a practically pure culture of the pneumococcus. The majority of the other cases of infection occurred during a brief reign of sepsis, due to the use of infected silkworm gut in the form of buried sutures. In about one half of the cases the suppuration was deep; in the other half superficial in character. A permanent good result, as far as the position of the uterus is concerned, was obtained in all but two of the suppurative cases—Case LXX and the case of pneumococcus infection mentioned above. In three patients (Cases I, XL, CXI) in whom extensive deep suppuration, with sloughing of the round ligaments and fascial edges, occurred in both wounds, the uterus remained in normal anteversion when last seen six years and a half, one month, and two months, respectively, after operation.

Diseases complicating Convalescence.—During convalescence, or, more accurately stated, during the first four weeks after operation, the following complications were noted:

	Cases
Typhus abdominalis	2
Malaria	10
Acute articular rheumatism	1
Acute articular rheumatism and pericarditis	1
Tertiary syphilis	1
Diphtheria	2
Acute croupous pneumonia	1
Acute catarrhal (ether) pneumonia	4
Acute right pleuritis	1
Tænia solium	1
Acute appendicitis	2
Acute cystitis from catheter infection	1
Sarcoma of left broad ligament	1

The patient who developed tertiary syphilis died of cerebral tumor three years and eleven months after operation. The rest all made good recoveries from their complications.

ULTIMATE RESULTS.

Failures.—Of the one hundred and fifteen cases tabulated, four were absolute and total failures. In one (Case XVI) the author failed to find the round ligament on the left side, no search being made for it on the right. Failures two and three (Cases LXX and LXXXIII) were due to sloughing of the round ligaments and of the edges of the fascial wound to which they were united. Failure four (Case XCIX) was the result of firm adhesions of the right tube and ovary, not recognized before operation, which pulled back the right round ligament into the abdomen within a month after operation. Two of these patients were subsequently cured, Case LXXXIII by vaginal, and Case XCIX by ventral fixation of the uterus.

There were five relative failures, due to giving way of one round ligament within the abdomen, the ligament in four cases being torn directly out of the uterus, in the fifth tearing at a distance of one centimetre from the uterus. In each of these five patients ventral fixation of the uterus was immediately substituted for the proposed shortening of the round ligaments, with resultant cure of the retroversion.

It will thus be seen that only two of the one hundred and fifteen patients were not cured of the conditions, in one prolapsus, in the other retroversion, for which operation was undertaken. One (Case XVI) declined a proposed ventral fixation; the second (Case LXX) died a sudden death from fatty degeneration of the heart five months after having the round ligaments shortened.

In speaking of ultimate results in the remaining one hundred and six cases, differentiation must be made between the anatomical result and the therapeutic result. To maintain the uterus in proper position is one thing, to cure your patient of her complaints quite another.

Anatomical Results.—The anatomical result in these one hundred and six cases was invariably all that could be desired, the uterus in each case remaining in normal anteversion at the date of the last obtainable examination. The appended table indicates the period of time after operation at which this final physical examination was made: 29 cases were last examined 1 month after operation; 12 in 2 months; 6 in 3 months; 1 in 4 months; 3 in 5 months; 5 in 6 months; 1 in 7 months; 2 in 8 months; 2 in 9 months; 2 in 12 months; 1 in 13 months; 1 in 14 months; 1 in 15 months; 1 in 16 months; 3 in 20 months; 1 in 21 months; 1 in 22 months; 1 in 23 months; 1 in 24 months; 1 in 26 months; 3 in 27 months; 1 in 28 months; 3 in 30 months; 2 in 31 months; 1 in 35 months; 2 in 36 months; 1 in 37 months; 1 in 38 months; 1 in 41 months; 2 in 42 months; 2 in 43 months; 1 in 47 months; 1 in 48 months; 1 in 50 months; 2 in 53 months; 1 in 55 months; 1 in 62 months; 1 in 66 months; 1 in 72 months; 1 in 73 months; 1 in 77 months; 1 in 79 months.

This makes an average period of observation of nearly seventeen months for each of the one hundred and six cases, a period long enough to warrant definite conclusions as to the permanency of the anatomical results of shortening the round ligaments. These perfect and lasting anatomical results are perhaps the more remarkable when it is considered that the majority of the women belong to the working classes, many of them being compelled to do housework of the hardest kind. In one patient (Case XXVII) the operation successfully stood the severe test of a subsequent complete acute prolapse of the uterus and vagina.

Therapeutic Results.—While the anatomical results may be thus readily and succinctly stated, an accurate report upon the therapeutic results becomes a much more complicated matter. The less searchingly we cross-question our patients, the more superficially we go into their after-histories, and the less thoroughly we examine them in the endeavor to obtain evidence to the contrary, the more likely are we to record a large number of therapeutic successes. But closer investigation often reveals one or more stumbling blocks to perfect satisfaction in the shape of complicating diseases present at the time of operation or developing soon thereafter, which complications more or less fully, more or less permanently, mar the therapeutic result.

The writer has considered it his duty to himself and to his patients to specifically and carefully investigate each case in which the anatomical result was perfect, while at the same time the therapeutic result was not all that could be desired. This was done with a view to finding, if possible, the explanation of the discrepancy.

Among conditions of disease present in patients at the time of operation, which shortening of the round ligaments and other operations combined at the same sitting could not be expected to relieve, and which persisted and caused their own symptoms after operation, the following were noted:

	Cases
Mild insanity	2
Chronic nephritis	2
Valvular lesions of the heart	6
Chronic appendicitis	1
Movable right kidney	13
Movable left kidney	1
Movable right and left kidneys	2
Bilateral salpingo-oöphoritis	1
Chronic metritis and right oöphoritis, with adhesions of ovary	1

Five of the cases of movable kidney were subsequently cured by nephropexy. In a sixth case nephropexy and shortening of the round ligaments were performed simultaneously. The case of chronic metritis and right oöphoritis with adhesions was cured by a subsequent pregnancy, which liberated the imprisoned ovary by carrying it out of the pelvis. The case of bilateral salpingo-oöphoritis was later radically cured by a vaginal bilateral salpingo-oöphorectomy performed by another surgeon. The remainder of these patients continued to suffer from the symptoms due to their respective complicating diseases. The complications, however, having been diagnosticated before operation, and the prognosis shaped accordingly, left no real ground for disappointment in these cases, the symptoms due to the retroversion being relieved in each of them.

In quite a number of cases, however, the development *after operation* of new morbid conditions caused disappointment, and called for repeated careful and critical analysis of the situation. Questioning and physical examination of the patient showed that the following diseases, developing *after* convalescence from the operation of shortening the round ligaments, had to a greater or less degree marred the therapeutic result:

	Cases
Progressive tertiary syphilis	1
Acute articular rheumatism	1
Chronic rheumatic endocarditis	1
Plumbism	1
Malaria	12
Acute dementia	1
Movable right kidney	7
Movable left kidney	1
Movable right and left kidneys	1
Chronic appendicitis	1
Acute perforative appendicitis	1
Endometritis	2
Salpingitis	1
Salpingo-oöphoritis	8
Fibromata uteri	1
Sarcoma of left broad ligament	1

Of these conditions developing after operation the following were surgically treated by the writer, all of them successfully:

Four of the movable kidneys by nephropexy.

The case of acute perforative appendicitis by removal of the appendix and drainage of abscess.

One case of endometritis and the case of salpingitis by curettage of the uterus.

One of the cases of salpingo-oöphoritis by unilateral and one by bilateral cœlio-salpingo-oöphorectomy.

The fibromata uteri case by bilateral salpingo-oöphorectomy.

The sarcoma of left broad ligament by extirpation of uterus, tubes, ovaries, and entire left broad ligament.

Among the conditions existing at the time of operation or developing thereafter, the various inflammatory conditions affecting tubes and ovaries, and movable kidney or kidneys, play so large and so important a rôle as to call for separate consideration.

Salpingo-oöphoritis in its Relations to Retrodeviations of the Uterus. —More or less endometritis exists in every case of retrodeviation of the uterus of some standing. This endometritis is not always, perhaps not even generally, cured by simply correcting the position of the uterus. Hence the rule to curette the uterus preliminary to any operation for the correction of the retrodisplacement. Endometritis predisposes to—indeed, tends directly—to the development of salpingitis, and, by continuation, of salpingo-oöphoritis. We have observed no less than eleven cases of the gradual development of endometritis, salpingitis, and salpingo-oöphoritis months after shortening the round ligaments, with resultant nullification or decided impairment of the therapeutic result of the operation.

Movablé Kidney in its Relations to Retrodeviations of the Uterus.— The frequent association of movable kidney or kidneys and retrodeviations of the uterus has been observed by the writer since 1890, and is drastically illustrated by the recorded observations made upon the one hundred and fifteen cases tabulated. No less than twenty-five patients are noted as suffering from movable kidney or kidneys, either existing at the time the round ligaments were shortened or developed after the operation. Of these twenty-five patients with movable kidney or kidneys, eleven were cured by right nephropexy and two by bilateral nephropexy. In three of these patients the nephropexy preceded the operation on the ligaments in point of time; in one case the two operations were performed simultaneously; while in nine instances nephropexy followed shortening of the round ligaments.

The therapeutic results may be summed up in the statement that the symptoms due to the positional deviations of the uterus and ovaries, for the correction of which shortening the round ligaments was undertaken, were permanently relieved in one hundred and eleven cases, the remaining four cases being complete failures.

The symptoms due to other pathological conditions coexisting in the patient were, except in a certain proportion of cases of endometritis and catarrhal salpingitis, not influenced by the operation. In other words, shortening of the round ligaments accomplishes therapeutically all that can logically be expected of it.

Hernia in Connection with shortening the Round Ligaments.—Only one patient (Case XIII) acquired hernia as the result of shortening the round ligaments. Both wounds of this patient healed by primary union. During the fifth month of her first pregnancy, nineteen months after shortening the round ligaments, right inguinal hernia developed. Soon after delivery, or about two years after operation, left inguinal hernia also appeared. Both herniæ are to be operated upon for radical cure after the termination of her present pregnancy, the second since operation.

Case XLII suddenly developed an acute strangulated right *femoral* hernia two years and nine months after, but certainly not as a result of, shortening the round ligaments. The hernia was successfully operated upon on the day of its occurrence, and remains radically cured to date.

Case XLVI had left inguinal hernia and retroversion of the uterus. Radical herniotomy was performed in conjunction with shortening the round ligaments. The patient remained cured of the hernia when last seen, sixteen months after operation.

Case LXI had retroversion and bilateral inguinal hernia, and double herniotomy was added to shortening the round ligaments. Suppuration and sloughing in both wounds was the result of repeated removal of her dressings by the mentally irresponsible patient. The left hernia recurred six months after operation; the right remained cured when last seen, nine months after operation.

Case LXXI presented a *right* inguinal and a *right* femoral hernia at the time of shortening the round ligaments. Both herniæ were operated upon with a view to radical cure at the same sitting with the ligament operation. When last seen, seventeen months after operation and two months after delivery at term, the inguinal hernia had returned, a new *left* femoral hernia had developed, while the original *right* femoral hernia remained cured.

In Case LXXIV two inguinal herniæ were operated upon for radical cure while shortening the round ligaments. Both herniæ remained cured when the patient was last seen, twenty-one months after operation.

To summarize: Two inguinal herniæ developed in one patient as a result of the operation; two *femoral* herniæ developed after, *but not as a result of,* the operation. Of seven herniæ existing at the time of operation, and operated upon for radical cure while shortening the round ligaments, four inguinal herniæ and one femoral remain cured; two inguinal herniæ have recurred. Four inguinal and two femoral herniæ are now present as against six inguinal and one femoral existing at the time of shortening the round ligaments. The two femoral herniæ stand in the relation of *post hoc,* but not *propter hoc,* to the operation of shortening the round ligaments. Two inguinal herniæ, both in the same patient, were produced, and four inguinal herniæ in four patients were cured by the operation of shortening the round ligaments. These hernia experiences all belong to the period at which the entire canal was opened and the round ligament was placed immediately behind the lips of the fascial wound, to which it was sutured. With the present improved technics, closing by a Bassini, the occurrence of an inguinal hernia after operation ought to prove a very great rarity, while the cure of herniæ existing at the time of shortening the round ligaments should become an almost absolute certainty. The inguinal canal can be much more securely closed in women than in men, owing to the necessity of providing for the spermatic cord in the latter.

Cases of Pregnancy following shortening of Round Ligaments.—Eighteen pregnancies have been observed, occurring in eleven of the one hundred and fifteen tabulated cases.

Of these eighteen pregnancies no less than six occurred in three of the five patients whose round ligaments parted within the abdomen during operation, and in whom ventral fixation of the uterus was substituted for shortening the round ligaments.

These six, therefore, are cases of pregnancy following ventral fixation of the uterus. One mother (Case XXXIX) died suddenly, from valvular disease of the heart, on the eve of confinement at term. Case XLIX had one miscarriage and two living children at term, both born without complications. The third patient (Case LXXIX) had one abortion at the fourth month and one difficult and disastrous labor at term—transverse presentation, inability to deliver, on account of high posterior position of cervix and obstruction formed by thickened anterior wall of uterus; rupture of uterus; cœlio-panhysterectomy; death from sepsis.

The records of the twelve pregnancies following shortening of the round ligaments are carried up to date of going to press:

Case XI: Three pregnancies, three living children at term.

Case XIII: Two pregnancies, two living children at term.

Case XX: One pregnancy, one living child at term.

Case XXXVIII: One pregnancy, one living child at term.

Case LXVII: One pregnancy, one living child at term.

Case LXXI: One pregnancy, one living child at term.

Case XLVI: Two pregnancies, one miscarriage at third month; pregnant seven months, without complications, when last seen.

Case XCVII: One twin pregnancy; miscarriage at third month.

There were thus ascertained twelve pregnancies in eight women whose round ligaments had been shortened. Of these twelve pregnancies two terminated in abortion, one was lost sight of after the seventh month, and nine ended with the safe delivery at term of living children. All the labors were easy and natural with one exception: Case LXXI had a transverse presentation, necessitating version and forceps to the after-coming head; mother well, child alive. In the seven patients whom it was possible to follow after each pregnancy, the uterus remains in anteversion.

SUMMARY.

Shortening the round ligaments is the only operation by which the retrodisplaced uterus can be brought into normal and physiological anteversion without establishing always pathological peritoneal adhesions. All other retroversion operations depend for their success upon more or less extensive, more or less firm, peritoneal adhesions.

Shortening the round ligaments, in capable hands, is as safe and as successful as the other retroversion operations.

Shortening the round ligaments is absolutely free from the disturbances and disasters of future pregnancy and parturition which are on record as having followed vaginal and ventral fixation of the uterus, its chief rivals.

Shortening the round ligaments is therefore indicated, and should be the operation of choice whenever and wherever it will meet the indications as well as, or better than, one of the rival procedures.

Shortening the round ligaments is indicated—

(a) In all uncomplicated cases of retroversion, retroflexion, and excessive mobility of the uterus requiring operative treatment.

(b) In extreme and aggravated cases of anteflexion of the uterus.

(c) In cases of retroverted, anteflexed uteri without adhesions.

(d) In simple prolapse of the ovaries when that condition calls for treatment.

(e) In cases of adherent retrodisplaced uteri, with or without adhesions of tubes and ovaries, these organs being otherwise in condition to call for their conservation. The adhesions are first to be severed by colpotomy, median cœliotomy, or an incision through the peritonæum at the internal ring.

Shortening the round ligaments does *not* compare in efficiency with ventral fixation of the uterus as a prolapsus operation.

Shortening the round ligaments should always be immediately preceded by curettage of the uterus. Other operations may be associated according to the indications in the particular case.

The round ligament is never absent. It may, however, after emerging from the internal inguinal ring, run an erratic course to an abnormal insertion (in the writer's experience in four per cent. of round ligaments).

Shortening the round ligament is best performed by opening the whole length of the anterior wall of the inguinal canal, drawing the ligament out at the internal ring, really shortening the intra-abdominal portion by stripping back the investing peritonæum and closing the wound after the manner of the Bassini operation for the radical cure of inguinal hernia, leaving and securing the shortened ligament in its natural habitat behind the lower edge of the internal oblique.

Of the author's one hundred and sixteen cases, four were absolute failures. In one of these the operator failed to find the round ligament on one side; failures two and three were due to sloughing of the ligaments; failure four to unrecognized adhesions of one ovary, which pulled back the round ligament of that side into the abdomen within a month. Two of these four patients were subsequently cured, one by vaginal and one by ventral fixation of the uterus. There were five relative failures, due to giving way of one round ligament within the abdomen. In each of the five patients ventral fixation was immediately substituted, with resultant cure of the retroversion. The writer preferred this course to trusting to one shortened ligament to hold up the uterus. One patient died within a week after operation of acute gangrenous appendicitis with septic peritonitis.

In the remaining one hundred and six patients the uterus remained in normal anteversion when last examined, the period of observation varying from one month to six years and a half after operation, and averaging over sixteen months for each of the one hundred and six cases.

The writer is convinced that these results can be improved upon.

Two inguinal herniæ, both in the same patient, occurred as a result of shortening the round ligaments. Six inguinal and one femoral hernia were operated upon simultaneously with shortening of the round ligaments. Of these two, inguinal herniæ recurred; the femoral and four inguinal herniæ remain cured.

Twelve pregnancies are known to have followed in eight of the successful cases. Of these, two terminated in abortion, one was lost sight of after the seventh month, and nine ended with the safe delivery at term of living children. In the seven patients whom it was possible to follow after each delivery, the uterus remains in anteversion.

The appended bibliography is as complete as the resources of the library of the New York Academy of Medicine enabled us to make it. The writer is deeply indebted to Dr. T. Alexander Lehman for valued aid in its preparation and in abstracting the literature.

BIBLIOGRAPHY.

1. ALQUIE: Memoire sur une nouvelle methode pour traiter les divers deplacements de la matrice. Bull. Acad. de med., Par., 1840-'41, t. vi, p. 223.
2. ALQUIE: Nouvelle methode pour traiter les divers deplacements de la matrice (Rap. de Villeneuve). Bull. Acad. de med., Par., 1844-'45, x, 192-195.
3. ARAN: Maladies de l'uterus. 1858, p. 1039.
4. FRITSCH, H.: Lageveraenderungen der Gebärmutter. Stuttgart, 1881, p. 136.

1882.

5. ALEXANDER, W.: A New Method of treating Inveterate and Troublesome Displacements of the Uterus. Med. Times and Gaz., Lond., 1882, i, 327.
6. ADAMS, J. A.: A New Operation for Uterine Displacements. Glasgow M. J., 1882, xvii, 437-446.

1883.

7. ALEXANDER, W.: A New Method of treating Displacements of the Uterus. Liverpool M.-Chir. J., 1883, iii, 113–124.

8. CAMPBELL, W. M.: Four Cases of Alexander's Operation for Prolapsus and Retroversion of the Uterus. Liverpool M.-Chir. J., 1883, iii, 235–240.

1884.

9. ALEXANDER, W.: The Treatment of Backward Displacements of the Uterus and of Prolapsus Uteri by the New Method of shortening the Round Ligaments. Lond., 1884, J. and A. Churchill, 71 p., 8°.

10. GARDNER, W.: Alexander's Operation on the Round Ligaments. Austral. M. J., Melbourne, 1884, n. s., vi, 390–396.

11. GARDNER, W.: The New Operation of shortening the Round Ligaments for Troublesome and Inveterate Displacements of the Uterus. Australas. M. Gaz., Sydney, 1883–'84, iii, 55–57.

12. GEHRUNG, E. A.: The Treatment of Backward Displacements of the Uterus and of Prolapsus by the New Method of shortening the Round Ligaments Critically Reviewed. Denver M. Times, 1884–'85, iv, 150–153.

13. LEDIARD, H. A.: Alexander and Adams' Operation on the Round Ligaments. Brit. M. J., London, 1884, i, 354.

14. MILIER, R.: Two Cases of Alexander-Adams' Operation for Displacements of the Uterus. Glasgow M. J., 1884 s., xxii, 121–125.

15. REID, W. L.: On the Operation (Alexander-Adams) of shortening the Round Ligaments for Uterine Displacements, with Three Cases. Brit. M. J., London, 1884, ii, 958.

16. WINSLOW, R.: A Case of Procidentia Uteri treated by shortening the Round Ligaments combined with Kolpo-perineorrhaphy. Med. News, Phila., 1884, xiv, 598.

1885.

17. ALEXANDER, W.: The Operation of correcting Some Uterine Displacements by shortening the Round Ligaments. Brit. Gynæc. J., London, 1885, i, 246–269.

18. ALEXANDER, W.: Shortening the Round Ligaments. Brit. M. J., Lond., 1885, ii, 671.

19. ALEXANDER, W.: On the Permanence of the Operation of shortening the Round Ligaments for Prolapse, with Hints as to its Mode of Performance. Med. Chron., Manchester, 1885, ii, 369–374.

20. ALEXANDER, W.: On the Operation of shortening the Round Ligaments. Edinb. M. J., 1884–'85, xxx, 1030–1033.

21. BEURNIER, L : Note sur l'anatomie des ligaments ronds au point de vue de l'operation d'Alexander. Union med., Par., 1885, 3, s., xi, 942.

22. DENEFFE: Raccourcissement des ligaments ronds pour la cure de la retroversion, de la retroflexion, et de la chute de l'uterus. Ann. Soc. de med. de Gand, 1885, lxiv, 135–138.

23. DOLERIS. A , et RICARD : Recherches anatomiques et operatoires a propos du raccourcissement des ligaments ronds, ou operation dite d'Alexander-Adams. Union med., Par., 1885, 3 s., xi, 865–869.

24. DURAND-FARDEL, R.: L'operation d'Alexander, raccourcissement des ligaments ronds, pour la cure de la retroversion et retroflexion de l'uterus. Gaz. med. de Par., 1885, 7 s., ii, 17–19.

25. ELDER, G.: A Case of Alexander's Operation of shortening the Round Ligaments in a Case of Aggravated Prolapse and Retroflexion. Obst. Gaz., Cincin., 1885, viii, 12–15.

26. HERMAN: Hypertrophy of Infravaginal Cervix; Prolapse; Alexander's Operation; Peritonitis; Death. Med. Times and Gaz., Lond., 1885, ii, 112.

27. IMLACH, F.: On shortening the Round Ligaments of the Uterus. Edinb. M. J., 1884–'85, xxx, 913–918.

28. MANGIAGALLI, L.: Le raccourcissement des ligaments ronds, d'apres la methode sanglante, comme traitement radical des deplacements de l'uterus en arriere et en bas. Gaz. hebd. d. sc. med. de Bordeaux, 1885, vi, 389, 399.

29. MASSE: Traitement des deviations uterines par le raccourcissement des ligaments ronds, operation d'Alquie, l'uteroinguinoraphie; operation d'A. Adams. Gaz. hebd. d. sc. med. de Bordeaux, 1885, vi, 95, 119, 138.

30. MUNDE, P. F.: Four Cases of Alexander's Operation of shortening the Round Ligaments of the Uterus for Retroversion. N. Eng. M. Month., Sandy Hook, Conn., 1884–'85, iv, 353–357.

31. NANCREDE, C. B.: Shortening of the Round Ligaments; Alexander's Operation. Med. and Surg. Reporter, Phila., 1885, iii, 681–683.

32. PARISH, W. H.: Alexander's Operation. N. Y. M. J., 1885, xii, 621.

33. POLK, W. M.: On Alexander's Operation for Retroversion of the Uterus. Phila. M. Times, 1884–'85, xv, 857–860.

34. RIVINGTON, W.: The Operation of shortening the Round Ligaments for remedying Uterine Displacements. Brit. M. J., Lond., 1885, i, 425.

35. SINCLAIR, A. J.: Notes on a Case of Alexander-Adams' Operation. Edinb. M. J., 1885–'86, xxxi, 250–253.

36. ZEISS: Ein Fall von Verkürzung der Ligamenta rotunda Uteri (Alexander-Adam'sche Operation). Centralbl. für Gynaek., Leipz., 1885, ix, 689–691.

37. DECAZE, P.: Operation d'Alexander. N. dict. de med. et chir. prat., Par., 1886, xi, 391–399.

1886.

38. DOLERIS et RICARD: Nouvelles etudes sur l'operation d'Alexander. Union med., Par., 1885–'86, 3, s., xi, 1057–1060.

39. DOLERIS: De l'operation du raccourcissement des ligaments ronds. N. arch. d'obst. et de gynec., Par., 1886, i, 10, 69, 158, 229.

40. FOWLER, G. R.: Two Cases of Alexander's Operation of shortening the Round Ligaments. Ann. Surg., St. Louis, 1886, iv, 42–46.

41. GARDNER, W.: Twenty Cases of Alexander-Adams' Operation. Austral. Med. J., Melbourne, 1886, viii, 437–445.

42. HARRINGTON, F. B. The Operation of shortening the Round Ligaments, with Two Cases. Boston M. and S. J., 1886, cxiv, 390–392.

43. HEYDENREICH, A.: L'operation d'Alexander. Semaine med. de Par., 1886, vi, 297.

44. REITH, S.: An Unsuccessful Case of Alexander's Operation. Jr. Edinb. Obst. Soc., 1885–'86, xi, 102–104.

45. MANRIQUE, I. E.: Étude sur l'operation d'Alexander, precedée de quelques considerations sur les deplacements de l'uterus, 4°, Paris, 1886.

46. MCDONALD, W. C.: Alexander's Operation. Homœop. J. Obst., N. Y., 1886, viii, 44–48.

47. POLK, W. M.: Alexander's Operation (shortening of the Round Ligaments); with a Report upon Fifteen Cases in Bellevue Hospital. Med. Rec., N. Y., 1886, xxx, 1–4.

48. POLK, W. M.: The Cure of Procidentia by Alexander's Operation. Am. J. Obst., N. Y., 1886, xix, 605.

49. POLK: A Successful Case of Alexander's Operation. Am. J. Obst., N. Y., 1886, xix, 158.

50. RIASENTSEV, L. H.: Abbreviatio ligamentorum rotundorum uteri (Alexander-Adams). Russ. Med., St. Petersb., 1886, iv, 209–212.

51. SLAVJANSKI: Zur Alexander-Adams'schen Operation. Verhandl. d. deutsch. Gesellsch. f. Gynäk., Leipzig, 1886, i, 257–265.

52. SMITH, A. L.: Remarks upon Alexander's Operation. Canada Med. Rec., Montreal, 1886–'87, xv, 25–29.

53. ZEISS: Zur Alexander-Adams'schen Operation. Verhandl. der deutsch.'Gesellsch. f. Gynäk., Leipz., 1886, i, 252–256.

1887.

54. ALEXANDER, W.: The Results of the Experience gained in Six and a Half Years of the Operation of shortening the Round Ligaments for Uterine Displacements. Tr. Internat. M. Cong., Wash., 1887, ii, 742–756.

55. ASHBY, J. A.: A Contribution to the Study of the Operation of shortening the Round Ligaments; Alexander's Operation. Obst. Gaz., Cincin., 1887, x, p. 57–65.

56. BOUILLY: Retroflexion de l'uterus à angle aigu; douleurs continuelles, operation d'Alexander; guerison. Bulletin et Mem. Soc., de chir. de Paris, 1887, n. s. xiii, pp. 134–139.

57. BYFORD: The Round Ligaments from a Case of Alexander's Operation. Journ. of the Am. Med. Assoc., 1887, p. 442.

58. CASATI, E.: Una modificazione all operazione. Med. Fogli., 1887, iv, 129, 193, 225.

59. COLLINS, J. W.: Shortening of the Round Ligaments in Retrodisplacements of the Uterus. Tr. Colorado M. Soc., Denver, 1887, pp. 84–89.

60. DUBREUIL, A.: Operation d'Alquié-Alexander. Gaz. hebd. soc. med. de Montpellier, 1887, ix, 385.

61. FOREMAN, J. The Alexander-Adams' Operation. Tr. Intercolon. M. Congr., Australas., 1887, Adelaide, 1888, i, 204–208.

62. GARDNER, W.: Alexander's Operation of shortening the Round Ligaments for Inveterate Displacements of the Uterus. Tr. Intercol. M. Cong., Australas, 1887, Adelaide, 1888, i, 214–225.

63. KELLOGG, J. H.: The Radical Cure of Retrodisplacements of the Uterus and Procidentia by Alexander's Operation and Median Colporrhaphy. Jr. Michig. M. Soc., Detroit, 1887, xi, 297–316, 3 pl.

64. KELLOGG, J. H.: Report of Twelve Cases of Alexander's Operation. J. Am. Med. Assoc., Chicago, 1887, ix, 225–231.

65. KELLOGG, J. H.: The Correction of Uterine Displacements by Alexander's Operation, with Report of Twenty Cases. Jr. Internat. Med. Cong., Wash., 1887, ii, 764–774.

66. MYNTER, H.: Alexander's Operation. Buffalo Med. and Surg. J., 6, 241–251.

67. POZZI, S.: Retroflexio uterine. Operation d'Alexander-Adams. Gaz. med. de Par., 1887, 7 s., iv, 121–124.

68. PUECH, P.: Quelques mots sur le raccourcissement des ligaments ronds. Montpel. med., 1887, 2 s., ix, pp. 245–257.

69. REID, W. L.: The Remote Results of the Operation of shortening the Round Ligaments for Displacements of the Uterus. Jr. Internat. M. Cong., Wash., 1887, ii, 757–764.

70. RIASENTSEFF, J.: Abbreviatio ligamentorum rotundorum uteri. 8°, St. Petersburg, 1887.

71. RIASENTSEFF: Abbreviatio ligamentorum rotundorum uteri. J. akush, i. jensk. boliez., St. Petersburg, 1887, i, 106–115.

72. SWIFT, J. B.: A Case of Alexander's Operation. Boston M. and Surg. J., 1887, p. 421.

73. THIRIAR, J.: Note, sur l'operation d'Alexander. Clinique Brux., 1887, i., 765–769.

74. WINSLOW, R.: Some Anatomical and Surgical Notes upon the Operation of shortening the Round Ligaments. Maryland Med. J., Baltim., 1887, xvii, p. 343–345.

1888.

75. BERRUTI, G.: Quarto e quinto caso di operazione di Alexander. Osservatore, Torino, 1888, xxxix, 793–796.

76. BEURNIER, L.: Etude sur les ligaments ronds de l'uterus et sur leur raccourcissement. (Operation d'Alexander.) Gaz. des hop., Par., 1888, xi, 233–240.

77. BROWN, E. J.: On the Application of Alexander's Operation to Procidentia Uteri, with Report of Two Cases. Med. Rec., N. Y., 1888, xxxiii, p. 240–242.

78. BURT: Cases of Alexander's Operation. Am. Gyn., Boston, 1887–'88, i, 142–144.

79. BYFORD: Lecture on the Operative Treatment of Retroversion. Alexander's Operation. J. of Am. Med. Assoc., Chicago, 1888, x, 349–352.

80. COE, H. C.: A Discussion on Alexander's Operation. N. Y. Med. J., 1888, xlvii, 297–301.

81. DOLERIS: Traitement des deplacements uterines; combinaison des operations plastiques avec raccourcissement des ligaments ronds, etude basee sur trente cas. Congr. franc. de chir., Par., 1888, iii, 625.

82. GALLI: Storia clinica di un caso di operazione di Alexander eseguita dal. Pr. f. G. F. Novaro, Osservatore Torino, 1888, xxxix, 361–366.

83. GARDNER, W.: Alexander's Operation of shortening the Round Ligaments of the Uterus. Transact. Intercol. Med. Congr., Austral., 1887, Adelaide, 1888, i, 214–225.

84. KELLOGG, J. H.: Report of Twenty-five Cases of Alexander's Operation. Ann. Gyn., Boston, 1887–'88, i, 107–115.

85. KELLOGG, J. H.: Report of Forty-eight Cases of Alexander's Operation. J. Am. Med. Assoc., Chicago, 1888, xi, 793–803.

86. KELLOGG, J. H.: Report of Sixty-three Cases of Alexander's Operation for shortening the Round Ligaments and Description of an Improved Method of Operation. Trans. Am. Ass. Obst. and Gyn., Phila., 1888, i, 222–225.

87. LEE, C. C.: The Value of Alexander's Operation in Cases of Complete Procidentia of the Womb. Trans. Med. Soc., N. Y., Syracuse, 1888, 405–410.

88. MORFSIII, A.: Un operazione di Alexander. Bull. d. Osp. di S. Casadi, Loreto, 1887–'88, i, 201–203.

89. MUNDE, P.: The Value of Alexander's Operation for shortening the Round Ligaments estimated from the Results of Seventy-three Cases. Am. J. Obst., N. Y., 1888, 1121–1138.

90. NAMMACK, W. H.: Complete Procidentia; Alexander's Operation; Cure. Med. Rec., N. Y., 1888, xxxiii, 300.

91. NEWMAN, II. P.: Alexander's Operation, with Report of Cases. Am. J. of Obst., N. Y., 1888, xxi, 1291–1302.

92. NOVARO, F., e BERRUTI, G.: Secondo e terzo caso di operazione di Alexander. Observatoire, Torino, 1888, xxxix, 769–771.

93. POLK, W. M.: Hysterorrhaphy and Alexander's Operation. Am. J. Obst., N. Y., 1888, xxi, 1271.

94. RODMAN, W. L.: Alexander's Operation with an Illustrative Case. Am. Pract. and News, Louisville, 1888, n. s., vi, 195–197.

95. ROUX: Sur l'operation d'Alexander-Adams. Revue med. de la Suisse rom , Geneve, 1888, viii, 645–656.

96. STRONG, C. P.: Six Cases of Uterine Displacements treated by shortening of the Round Ligaments with Remote Results. Boston Med. and Surg. J., 1888, cxviii, 166–168.

97. VARNALY, L.: Operatiunea lui Alexander, in prolapsue si retroversiunea uterulin. Bucuresci, 1888, 86.

1889.

98. BAKER, C. II.: Report of Two Cases of Alexander's Operation, One for Anteversion. Med. Bull., Phila., 1889, xi, 209.

99. BERRUTI, G.: Sull operazione di Alexander negli spostamenti dell utero. Considerazioni cliniche ed anatomo-patologiche sopra cinque casi operati nell' ospedale Maria Vittoria. Osservatore, Torino, 1889, xi, pp. 25, 97.

100. CORBETT, C. N.: Notes on Alexander's Operation, Illustrated by Six Cases. New Zealand Med. J., Dunedin, 1888–'89, ii, 208–218.

101. DEBIERRE et DUTILLEUL PELTIER: Note sur les ligaments ronds de l'uterus et l'operation d'Alexander. Bull. med. du nord, Lille, 1889, xxviii, 102–107.

102. DOLERIS: Raccourcissement des ligaments ronds ; modification du manuel operatoire consistant dans la reunion et la suture des extremites des ligaments. N. arch. obst. et de gynéc., Paris, 1889, iv, 49–54.

103. GAIRAL, PERE: Reflexions sur les discussions qui ont eu lieu a la société de chirurgie au sujet du traitement des diverses variations et descentes de l'uterus par l'operation d'Alquie-Alexander. Gaz. de gynéc., Paris, 1889, iv, 145–148.

104. GILLIAM, D. J.: Report on Cases of the Alexander Operation. Times and Reg., Phila., 1889, xx, p. 222.

105. HILLS, A. J.: Notes on Alexander's Operation. N. Y. M. Times, 1888–'89, xvi, 271.

106. KELLOGG, J. H. : Report of Seventy-three Cases of Alexander's Operation for shortening the Round Ligaments and Description of an Improved Method of Operation. Transact. Mich. Med. Soc., Detroit, 1889, xiii, 311–332.

107. OSTRCM, H. J. : Shortening the Round Ligaments for Displacement of the Uterus. (Alexander's Operation.) Homœopath. J. Obst., N. Y., 1889, xi, 9–75. ·

108. POTHERAT, E. : Operation d'Alexander Gaz. méd. de Paris, 1889, 7, s., vi, 29, 40.

109. RICHELOT, L. G. : Sur l'operation d'Alexander (raccourcissement des ligaments ronds). Bull. et mem. soc. de chir. de Par., 1889, n. s., xv, 268–272.

110. TERILLON, O. : Rétroversion et rétroflexion de l'uterus ; operation d'Alexander. In his Leçons de clin. chir., 8°, Par., 1889, 97–110.

111. TRELAT : De l'operation d'Alquie-Alexander ; raccourcissement des ligaments ronds. Ann. de gynec., Par., 1889, xxi, 161–180.

112. SCHWARTZ, F. : Du traitement des deplacements et des deviations uterines par le raccourcissement des ligaments ronds ; operation d'Alquié-Alexander. Bull. et mem. soc. de chir. de Paris, 1889, n. s., xv, 241–255.

113. SEGOND : Chute de l'uterus et cystocele ; raccourcissement de ligaments ronds et elytrorrhaphie. Gaz. des hop., Par., 1889, xx, 313.

114. SHROPSHIRE, L. L. : A Case of Alexander's Operation. Trans. Texas Med. Assoc., Austin, 1889, 216–218.

115. STONE, I. S. : Alexander's Operation ; Hysterorrhaphy. Gaillard's Med. J., 1889, xlix, 242.

116. STONE, I. S. : A Case of Prolapsus Uteri treated by Amputation of the Cervix and Partial Alexander's Operation. Virginia M. Advance, Warrenton, 1889, I, I.

1890.

117. ALLOWAY, T. J. : A Report of Twenty Cases of shortening the Round Ligaments at the External Inguinal Opening in the Treatment of Retrodisplacements of the Uterus. Montreal Med. J., 1889–'90, 721–730.

118. BASTIANELLI, R. : Sull operazione di Alexander modificata e sulla laparoisteropessia. Estratto dal Bulletino della Società Canciscana, fasc. ii, 1890, Seduta, viii, April 11, 1890, Roma, 1890.

119. BOMPIANI, A. : Delle operazioni cruenta per cura della retroflessione dell' utero e specialmente delle operazioni di Alexander e di Olshausen. Ann. di ostet., Milano, 1890, xii, 433–450.

120. CARPENTER : Alexander's Operation with a New Method for securing the Round Ligaments. Journal Am. Med. Assoc., Chicago, 1890, xxiv, p. 308.

121. CUGIANI, L. : Resoconto di cinque operazioni d'Alexander. Ann. di ostet., Milano, 1890, xii, 359–371.

122. DRESSLER, OSKAR : Ein Beitrag zur Beurtheilung der Alexander-Adams-'schen Operation. Kiel, 1890, L. Handorff, 34, 8°.

123. EDEBOHLS, G. M. : A Modified Alexander-Adams Operation. N. Y. Med. J., 1895, lii, 400–404.

124. FRY, H. D. : The Alexander Operation ; Report of a Case. Virginia Med. Monthly, Richmond, 1890–'91, xvii, 165–171.

125. HAGE, I. L. : De verkorting der ligamenta rotunda bij liggingsafwijkingen van den uterus. 8°. Leiden, 1890.

126. LAGRANGE : Opération d'Alexander pour rétroversion uterine ancienne. Guérison. Bull. et mem. soc. de chir. de Paris, 1890, n. s., xvi, 183.

127. LIMONT, I. : Notes of Three Cases of Alexander's Operation for shortening the Round Ligaments. Rep. Proc. Northumb. and Durham M. Soc., Newcastle upon-Tyne, 1890-'91, 61-63.

1891.

128. ALEXANDER, W. : Reciprocal Effects of Pregnancy and Parturition upon the Operation of shortening the Round Ligaments of the Uterus. Brit. Med. J., Lond., 1891, i, 348-350.

129. BRANHAM : Shortening the Round Ligaments of the Uterus. Trans. Am. Ass. Obst. and Gyn., Phila., 1890, iii, 89.

130. EDEBOHLS, G. M. : The Remote Results of shortening the Round Ligaments for Uterine Displacements by the New or Direct Method. Am. J. Obst.. 1891, xxiv, 582-586.

131. JOHNSON, F. W. : Eleven Alexander-Adams' Operations. Boston M. and S. J., 1891, 404-406.

132. NEWMAN, H. P. : The Remote Results of shortening the Round Ligaments for Uterine Displacements by the New or Direct Method. Am. J. Obst., N. Y., 1891, 257.

133. RICHMOND, J. W. : Alexander's Operation ; Cystitis. Jr. Med. Assoc. Missouri, Kansas City, 1891, 28-30.

1892.

134. BLAKE : Alexander's Operation. Boston Med. and Surg. J., 1891, cxxv, p. 622.

135. BLEYNIE, L. : Opération d'Alquié-Alexander. Limousin méd., Limoges, 1892, xvi, 163-165.

136. CHALOT : Nouvelle méthode de raccourcissement des ligaments rondes de l'utérus. Ass. franç. pour l'avanc. de science. Sess. de Pau ; Séance, 17, Sept., 1892. Ref. Rép. univ. d'obst. et de gyn., Par., 1892, vii, 444, 445.

137. CITTADINI : Des traitements des déplacements de l'utérus ; rétro-déviations et prolapsus ; dix cas d'hystéropexie ligamentaire on raccourcissement des ligament ronds. Journ. de med. chir., et pharm. Ann. Bruxelles, 1892, i, 15-91.

138. JOHNSON, F. W. : Two Cases of Pregnancy and Labor at Term following the Alexander-Adams' Operation. Boston Med. and Surg. J., 1892, cxxvi, 363.

139. MORET : Métrites, prolapsus utérins et rétroversion chez une vierge ; curettage, opération d'Alexander, crises nerveuses et stercoremie. Union med. du nord-est, Reims, 1892, xvi, 47-49.

140. NITOT : Quelques considérations à propos de l'opération d'Alexander. Rev. obst. et gyn., Par., 1892, viii, 228-234.

141. POZZI, S. : Traité de gynécologie. Deuxième édition, Paris, 1892, p. 487.

1893.

142. CITTADINI : Complications opératoires du raccourcissement des ligaments ronds. Soc. de gyn. et d'obst. de Bruxelles. Séance 30, avril, Rép. univ d'obst. et de gynéc., Paris, viii, 4€5.

143. DULET : Modification de l'opération d'Alexander. Gaz. d. h'p. de Toulouse, 1893, vii, 268.

144. GELPKE, L.: Beitrag zur operativen Behandlung der Lageveränderungen der Gebärmutter. Zeitschr. f. Geb. u. Gyn., Stuttgart, xxvi, 335–350.

145. RUMMEL, E.: Observations anatomiques et cliniques concernant l'opération d'Alexander. Rev. méd. de la Suisse roman., Genève, vol. xiii, 470–484.

146. LANZ: Die Alquié-Alexander'sche operation zur Beseitigung des Prolapsus und der Retroflexio uteri nach Kocher. Arch. f. Gyn., Berlin, lxiv, 348–380.

147. LUDLAM: Hysterorrhaphy and Tubo-ovariotomy after Failure of Alexander's Operation. Clinique, Chicago, xvi, 329.

148. MUNDÉ, P. F.: Results of Alexander's Operation. Int. Clin., Phila., 1893, iii, 263.

149. SECCHI, E.: Contributo alla cura della retroversione uterina mediante l'operazione d'Alexander. Atti. d. Ass. med. lomb., Milano, 1893, 55–82.

150. WARDE, J.: De l'intervention chirurgicale dans les retrodéviations de l'uterus. Paris, Ollier-Henry, Thèse, 64.

1894.

151. BATCHELOR, F. C.: Notes on Alexander's Operation for shortening the Round Ligaments. N. Zealand M. J., Dunedin, vii, 226–229.

152. BIRD, A.: Alexander's Operation. N. Y. J. of Gyn. and Obst., i, 37.

153. BLAKE, J. G.: An Attempt to perform an Alexander Operation in a Case in which the Round Ligament had been removed in a Previous Laparotomy. Boston M. and S. J., 1894, cxxxi, 611.

154. BYFORD, H. T.: Prolapse of the Uterus; Alexander's Operation. Internat. Clin., Phila., vol. ii, 248–252.

155. CASATI, E.: Modificazione all' operazione dell' Alexander nella cura delle antiflessioni dell' utero   Gazz. med. lomb., Milano, 1894, 333.

156. CASATI, E.: Una modificazione all' operazione di Alexander. Atti dell' Acad. delle Sc. Mediche di Ferrara, iii.

157. CITTADINI: Remarques cliniques et operatoires sur quinze cas de raccourcissement des ligaments ronds. Congr. périod, internat. de gynec. et d'obst., 1892, Bruxelles, 717–729.

158. DAVENPORT, F. H.: Ultimate Results of Treatment of Backward Displacements of the Uterus by Pessary, with Special Reference to the Alexander-Adams Operation. Am. J. Obst., N. Y., 1894, xxx, 91–99.

159. EDEBOHLS, G. M.: Discussion. Trans. Am. Gyn. Soc., 1894, p. 218.

160. EDGE, F. A.: A Case of Kocher's Radical Modification of the Alexander-Alquié Operation. Lancet, London, i, 146.

161. KELLOGG, J. H.: Shortening the Round Ligaments vs. Ventral Fixation for Retroversion. Mod. Med. and Bacteriol. Review, Battle Creek, 1894.

162. LAMORT, R.: De l'influence comparée du raccourcissement des ligaments ronds et de l'hysteropexie abdominale au point de vue obstétrical. Bordeaux, 1894 118 p., 4°, No. 49.

163. MAGNOL, E.: Du traitement des déviations de l'utérus par l'opération d'Alquié, dite Alquié-Alexander. N. Montpél. méd., iii, 209, 735.

164. MOREAU, C.: Du raccourcissement des ligaments ronds appliqué à la guérison des déplacements de la matrice. Brux., 1894, F. Hazez., 90 p., 8°. (In Mém. couron Acad. roy. de méd. de Belg., Brux., 1894, xiii.)

165. MUNDÉ, P. F. : Ten Years' Experience with Alexander's Operation for shortening the Round Ligaments of the Uterus. M. Rec., N. Y., xlvi, 33.

166. NEWMAN, H. P. : Six Years' Experience in shortening the Round Ligaments for Uterine Displacements. Ann. of Gyn. a. Pæd., Phila., vol. vii, 625-636.

167. ROBSON, A. W. M. : Inveterate Retroflexion of the Uterus treated by the Alexander-Alquié Operation or by Hysteropexia. Quart. M. J., Sheffield, iii, 40-43.

168. SOUPART : Rapport de la commission qui a été chargee de l'examen du memoire de M. de Dr. Camille Moreau à Charleroi, intitulé : Du raccourcissement des ligaments ronds appliqué à la guérison des déplacements de la matrice. Bull. acad. roy. de méd. de Belge, Brux., 313.

169. STARK, M. E. : Clinical Studies in Alexander's Operation. Women's M. J., Toledo, 1894, iii, 57-60.

170. WARFRINGE : Sur le traitement du prolapsus uterin par le raccourcissement opératoire des ligaments utérins ronds (C.-D. Josephson). Gaz. de gynéc., Par. 1894, ix, 343-345.

171. WERTH, R. : Ueber die Anzeigen zur operativen Behandlung der Retroflexio uteri mobilis, nebst einem Beitrag zur Würdigung der Alexander'schen Operation. Festschrift zur Feier des 50 Jähr. Jubiläums, etc., Wien, 53-114.

1895.

172. BLAKE, J. G. : The Ultimate Results after Alexander's Operation. Med. and Surg. Rep., Bost. City Hosp., 1895, 6 s., 87-94.

173. CLEGHORN, G. : Alexander's Operation for Retrodisplacements of the Uterus. N. Zealand M. J., Dunedin, 1895, viii, 234-243.

174. CLEVELAND, C. : The Alexander Operation. Am. Gynæc. and Obst. J., N. Y., 1895, vi, 704-725 (Discussion), 857-867.

175. EDEBOHLS, G. M. : Discussion. Trans. Am. Gyn. Soc., 1895, p. 153.

176. FABRICIUS, J. : Zur Technik der Alexander-Alquié'schen Operation. Centralb. f. Gynaek., Leipz., 1895, xix, 786-790.

177. FRANK, J. : A Ligament Carrier for Alexander's Operation. Am. J. Obst., N. Y., 1895, xxxii, 280-282.

178. KUESTNER, O. : Alexander's Operation. Centralblatt f. Gynaek., Leipzig, 1895, xxi, 177-180.

179. KUMMEL, E. : Ueber Endresultate der Alexander'schen Operation. Centralb. f. Gynaek., Leipz., 1895, xix, 355-357.

180. RONCAGLIA, G. : Una parola in favore dell' operazione d'Alexander. Rassegna di sc. med. Modena, 1895, x, 263-275.

181. SMITH, A. L. : My Experience with Ventrofixation and Alexander's Operation. Am. Gynæc. and Obst. J., N. Y., 1896, vi, 852-857.

182. STOCKER, S. : Zur Alexander Operation. Cor. Bl. f. schweiz. Aerzt., Basel, 1895, xxv, 769-773.

1896.

183. ADAMS, J. A. : On shortening of the Round Ligaments in Displacements of the Uterus. Glasgow M. J., 1896, xiv, 435-444.

184. BONNET, S. : Des indications de l'opération d'Alexander. Semaine gynéc., Par., 1896, i, 58-60.

185. CLARKE, A. P. : A Consideration of the Value of Alexander's Operation

compared with that by Anterior Fixation of the Uterus. J. Am. M. Ass., Chicago, 1896, xxvi, 515.

186. EDEBOHLS, G. M.: The Indications for Ventral Fixation of the Uterus. Medical News, Phila., March 14, 1896.

187. EDEBOHLS, G. M. : Discussion on the Alexander Operation. Trans. N. Y Obst. Soc. Reported in Am. Gyn. and Obst. Jour., April, 1896, pp. 534–538.

188. JOHNSON, F. W.: The Alexander Operation. Am. Gynæc. and Obst. J . N. Y., 1896, viii, 457–467 (Discussion), 534–543.

189. MARTIN, F. H. Alexander's Operation without Buried Sutures. Am Gynæc. and Obst. J., N. Y., 1896, viii, 468.

190. McGANNON, M. C.: Extraperitoneal shortening of the Round Ligaments. (The Alexander-Adams Operation.) Am. Gynæc. and Obst. Journ., August, 1896, p. 195.

191. MUNDÉ, P. F. : The Indications for Alexander's Operation. Med. News, N. Y., 1896, lxviii, 281.

192. NITOT: À propos de la rétroversion ; sur un nouveau point de repère de visu destinée a faciliter la recherche des ligaments ronds dans l'opération d'Alexander. Rev. obst. et gynéc., Par., 1896, xii, 14–22.

193. NOBLE, C. P. ; Suspensio Uteri with Reference to its Influence upon Pregnancy and Labor. Am. J. Obst., August, 1896, p. 161.

194. STOCKER, S.: Ueber den Einfluss der Alexander Operation auf die Geburt einerseits und die Wirkung von Schwangerschaft und Geburt auf die Alexander Operation anderseits. Centralbl. f. Gynaek., Leipz., 1896, xx, 550–554.

59 WEST FORTY-NINTH STREET, NEW YORK.